Advances of Relativity and Cosmology

The only way to progress physics is to step out of the known paradigm and see its mistakes.

A.S. Šorli
Bijective Physics Institute
Slovenia

Content:

1. **Einstein's misunderstanding of time**

2. **Time-invariant space and cosmology**

3. **Advances of Relativity Theory**

4. **Time-invariant Superfluid Quantum Space as the Unified Field Theory**

5. **Schwarzschild energy density of superfluid quantum space and mechanism of AGNs' jets**

6. **Multiverse in permanent Equilibrium**

7. **Cosmo-biology: Universe, Life and Consciousness**

8. **Mass Gap Problem and Planck Constant**

Einstein's misunderstanding of time

Abstract

Einstein has kept time as the dimension of the space-time continuum that is supposed to be a fundamental arena of the universe. My research confirms time is the duration of changes, i.e., motion run in the time-invariant universal space that has Euclidean shape, it is infinite. Black holes in the centre of galaxies are rejuvenating systems of the universe. In these black holes old matter is transforming back into the fresh energy of elementary that AGNs are throwing in the intergalactic space in the form of huge jests. These jets are fresh material for new star formation. The universal process of continuous rejuvenation is eternal.

Key words: space, time, cosmology, Olbers' paradox, presentism, eternalism.

1. Introduction

The model of space-time as the fundamental arena of the universe is being replaced by the model of the time-invariant universal space, where time is merely the duration of change, i.e., motion (Šorli, Čelan, 2020). In time-invariant universal space time is not its 4th dimension. Experimental physics confirms this view, with clocks we measure the duration of a material change, i.e., motion in space. Moreover, the rigorous analysis of Special Relativity formalism of the fourth coordinate of space-time confirms that the fourth coordinate X4 is not time *t*:

$$X_4 = ict \quad (1),$$

In Eq. (1) *i* is the imaginary unit, *c* is the light speed, and *t* is duration. In Eq. (2) the time *t* is the duration of photon motion in space, time *t* is not X_4 (Fiscaletti, Sorli, 2017)

$$t \neq X_4 \quad (2) \, [2].$$

Einstein has interpreted the time *t* as the 4th coordinate X_4 of a Minkowski manifold. He wrote: "If we replace x, y, z, $\sqrt{-1}\,ct$ by x_1, x_2, x_3, x_4, we also obtain the result that $ds^2 = dx_1^2 + dx_2^2 + dx_3^2 + dx_4^2$ is independent of the choice of the body of reference. We call the magnitude *ds* the "distance" apart of two events or four-dimensional points. Thus, if we choose as time variable the imaginary variable $\sqrt{-1}\,ct$ instead of the real quantity *t*, we can regard the continuum space-time, in accordance with the special theory of relativity, as an "Euclidean" four-dimensional continuum, a result following by the consideration of the preceding section" (Einstein, 1916). In the above citation, Einstein suggests that we can choose the time variable *t* as the imaginary variable, can be written as follows:

$$t = \sqrt{-1}\,ct \quad (3).$$

Eq. (3) is false because on the left side of the equation we have t and on the left side we have $\sqrt{-1}\,ct$. Combining Eq. (3) with equation well known equation $X_4 = ict$ we get:

$$X_4 = itc^2\sqrt{-1} \qquad (4).$$

Einstein did a mistake keeping and interpreting time as the dimension of a four-dimensional continuum. Physics is still today suffering this misinterpretation of time that is solved in this article: time is the duration of material change, i.e., motion in time-invariant space.

Considering time as the duration of a change running in space, the universal space results as being time-invariant; the duration of a given event running in the universal space does not change in any way the physical properties of space and is not a part of the space. NASA has measured in 2014 that the universe space has Euclidean shape, measuring the angles between three stellar objects and getting 180° with 0.4% margin of error. This means that the universe has Euclidean shape and is infinite (NASA, 2014).

The idea of time-invariant structure of the universe was recently presented also by Hans J. Farr and Michael Hey: "One more cosmological possibility might perhaps need to be considered here, namely that the hierarchical structuring of masses in the universe which was considered in the above calculation could perhaps also be a time-invariant cosmic structuring, meaning that even though the universe undergoes an expansion in cosmic time, its hierarchical structuring endures or persists. Of course an expanding hierarchical universe must also change its mass density, however in such a way that the hierarchical structuring of matter persists, i.e. a time-invariant scale-invariance under these auspices must be considered" (Fahr, Heyl, 2020). The right understanding of time as the duration of change, i.e., motion in time-invariant space is in our view one of the most important elements of 21st-century physics progress.

2. Universe is time-invariant and eternal

Today's cosmology examines the universe from the perspective of the universe is existing in some linear time that has physical existence. We are seeing the universe as something that has started long ago and is still developing in the present day. A rigorous examination of what is time confirms that time as the duration enters existence only when measured by the observer. There is no physical time running in the physical universe on its own. Universal changes are irreversible. When change X+1 enters existence, change X is not in existence anymore. When change X+2 enters existence, change X+1 is not in existence anymore. Changes run in a time-invariant universal space where there is no past, there is no present and there is no future. The linear time "past-present-future" exists only in the human brain, it has its physical origin in neuronal activity (Šorli, Čelan, 2020).

The only universe that exists is the one we can observe and measure. NASA has measured universal space has a Euclidean shape and is infinite. We are living in an infinite time-invariant universe where there is no physical past and there is no physical future. From this perspective, it makes no sense to build a hypothesis about the begging of the universe in some remote physical past because such a past is non-existent. Would be more opportune to build cosmology only on the basis of astronomical observation and without hypothetical speculations about some remote beginning in some remote physical past; the universe is time-invariant which means eternal. There was no beginning and there will be no end. Active galactic nucleus (AGN) in the centre of galaxies are transforming old matter into fresh energy in the form of elementary particles. These particles are forming huge jets that are thrown out in the intergalactic space. Black holes in the centre of galaxies are rejuvenating systems of the universe that is eternal without the beginning and without an end (Šorli, 2020).

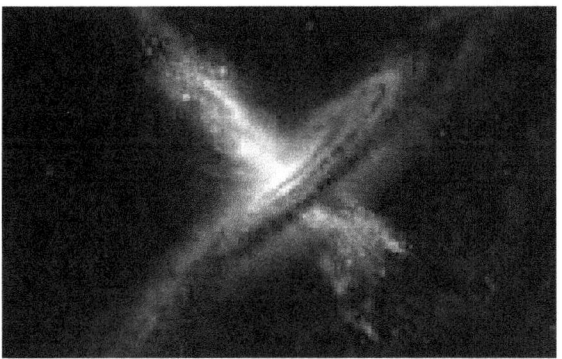

Figure 1: Jets from the black hole in the centre of galaxy
(with permission of European Southern Observatory - ESO)

The universe is eternal and this eternity is NOW. Humans, we experience the time-invariant nature of the universal space as NOW. Eternity is not extending infinitely back into the past and is not extending infinitely ahead in the future. Past, present and future exist only in the human mind; eternity is NOW (Fiscaletti, Sorli, 2014). Intuitively Einstein knew this and he expressed this in the following famous words: "People like us who believe in physics know that the distinction between past, present, and future is only a stubbornly persistent illusion". Still, in physics, he kept time as the fourth dimension of the model of the space-time continuum. The space-time continuum model has no correspondence in the physical world where we observe only material changes running in time-invariant space. Time as duration enters the existence when we me measure it.

3. Olbers' paradox solution

The formalism for the luminosity of a star at a finite distance from the Earth is as follows:

$$L = b \cdot 4\pi \cdot d^2 \quad (5),$$

where b is the apparent brightness of the star, L is its luminosity, and d the distance to the star. As a solution for Olbers' paradox, I argue in this way: the luminosity of stars which are in the area of finite distance from us is too low to make a day when we have a night. The light from the stars which are at infinite distance will never reach us and so cannot make a day when we have a night.

Harisson definition of the Olbers' paradox is as follows: "Olbers in 1826 was the first to show that the radiation density everywhere in an infinite static universe should equal to the radiation density at the surface of the stars. Hence, Olbers' paradox is that the sky is dark at night" (Harisson, 1964).

Olbers did not take into account that only the light from stars that are at a finite distance from the Earth can reach us. The light of stars that are infinitely far away cannot reach us, and that of stars at a finite distance is not strong enough to make the night a day. This is also the conclusion of Harrison article back in 1964: "Thus the radiation-level is low in our universe at present and the night-sky is dark simply because the stars are so widely separated from each other, or, in other words, because the characteristic time τ_0 is so very large (Harisson, 1964).

Our comment on the last Harrison's citation is that time in the solution of Olbers' paradox does not count. Time is the duration of light motion from the stars to the Earth. Time here does not play any role in Olbers' paradox solution. The decisive role is played by the distance and the luminosity of stars; the luminosity of stars at a finite distance is not strong enough to make a night sky a day. The light from stars that are at an infinite distance will never reach us and does not count.

Sakar's and Jeffries's research suggests that the paradox is resolved by the fact that the universe is expanding, which means that distant light has not yet reached us (Sarkar, Jeffries, 1997). Taking into account that the universe is expanding, our solution is still valid, namely, that the light of stars which are infinitely way from us will never reach us and the light of stars on a finite distance is not strong enough to make of a night sky a day.

Knutsen's research suggests: "It is dark at night because: (1) the speed of light is finite; (2) the universe is still young, and the stars only started to shine rather recently; (3) the light sources in the universe are small; (4) the light sources in the universe are far apart, so the energy density in the universe is very small. Both obscuration and expansion are quite unimportant" (Knutsen , 1997). None of these conclusions above is an exact explanation. It is dark at night because the light coming from the stars at a finite distance is not strong enough to make a night sky a day.

Harari's research suggests: "For terrestrial observation, when we consider the resolution of Olbers' paradox, we can use the apparent magnitude 8 as a conservative limiting magnitude. Thus, we can restate our earlier conclusion by noting that any star that has an apparent magnitude of fainter than 8 will not be visible to the average unaided human eye, and can, therefore, be considered non-existent for the analysis of the cumulative apparent brightness of the night sky under observation by the unaided eye. This limiting apparent magnitude constraint acts as a high-pass apparent brightness filter that effectively removes all fainter stars from the population of visible stars as far as unaided eye vision is concerned, thereby rendering finite the population of visible stars, irrespective of the initial size of the population of stars that exist in the universe, a population that may indeed be infinite" **(Harari, 2019).**

Harari's solution of Olber's paradox is tuned with the solution proposed in this article: the stars in the universe that have an apparent magnitude below 8 and are at a finite distance from the Earth are invisible to the unaided human eye, which means that the luminosity of stars at a finite distance is not strong enough, to make the night sky a day.

Olber's solution is deeply related to tour understanding of infinity. Infinite distance plus 1000km still is an infinite distance. "infinite distance" is not a metrical term that would describe some distance from our position in space to some physical object far away in the space. From the mathematical point of view, this is the problem of infinite numbers posed by German mathematician Georg Cantor: is an infinity of real numbers bigger than an infinity of natural numbers? Slovenia mathematician Ivan Vidav solved this problem proving that if we say that the infinity of real numbers is bigger than the infinity of natural numbers, there is no contradiction (Vidav 1959). If we say that both infinities are equal there also there is no contradiction. This clearly confirms that infinity is not a metrical term. In this perspective human imagination of the size of the universe is limited. We do not know exactly what infinity means, but one thing is clear: no light from the stars that are in infinity can reach us.

4. The time-invariant space model surpasses presentism and externalism

Presentism believes that past and future are somehow coexisting in the present moment that is the only moment that exists. The entire history and the entire future of the universe are squished in the present moment. Eternalism believes that time is extending infinitely far into the past and infinitely far into the future. We can imagine presentism as the mathematical point and eternalism as the infinite straight line.

We will examine both views from the pragmatic view. You take a stone, you keep it above your leg, and then you throw it on your leg. When the stone hits you on the leg, you feel the pain. Before you lift the stone from the ground there was no pain in your leg. This proves that in the present moment events are not coexisting. They are following each order in the sequential order: 1. lifting the stone, 2. keeping the stone above the leg, 3. throwing the stone on the leg. We have seen in chapter 2 that sequential events in the universe are irreversible and are not coexisting. Presentism seems wrong; it is not the truth that everything coexists in the present moment. Eternalism sees the event with the stone is happening in the linear physical time one after other. Eternalism keeps the past as something real despite nobody ever reach into the past. For eternalism, the universe runs in some linear time that nobody ever measured and observed. Seems this view is not right.

In cosmology, presentism is an inspiration for the "block universe" model where everything that happens is coexisting. Eternalism is the inspiration for the "space-time continuum" where we can have "closed time-lines" discovered by Kurt Gödel. The closed time-lines theoretically allow one could travel back in time and kill his grandfather and so it could not be born. That's why Gödel said: "In any universe described by the Theory of Relativity, time cannot exist" (Sorli, Fiscaletti, Gregl, 2013). Gödel's discovery is still today interpreted wrong by some researchers who think that his development of General Relativity equations and consequently the discovery of closed time-lines indicates that one could travel back in time. On the contrary, Gödel was strictly against the idea of time travel. One can travel only in time-invariant space where there is no present, no past, and no future.

In the time-invariant space model, there is no physical time in the sense of present time as considered in presentism and there is no time in the sense of some linear physical time as considered in eternalism. Time as duration enters existence when measured by the observer. Time-invariant space that we humans experience as the present moment is the fundamental non-created eternal background of the universe. Universal changes run in this time-invariant space that is eternity itself. Humans experience the flow of changes in the frame of psychological time "past-present-future" and that's why we experience changes running in some linear physical time that is not there. We are "projecting" our psychological linear time that is the product of neuronal activity in the physical universe (Šorli, Čelan, 2020).

In the universe, we can only observe the relative rate of clocks and not some "relative time". Clocks run on the GPS satellites for 45 microseconds per day faster than on the Earth surface because of the General Relativity effect. And they run slower for 7 microseconds per day because of the SR effect (Neil, 2002). Clocks tick only in time-invariant space and not in some physical time. What is "relative" in the universe is not time, relative is the rate of clocks and velocity of material changes in general. A twin on the Moon would age faster than his brother on the Earth's surface because the velocity of changes on the Moon is a bit faster regarding the Earth's surface. The weaker is gravity faster is the rate of clocks and aging too. In interstellar space where gravity is weak, the twin would age a bit faster than his brother on the Moon surface or on Earth surface. But there is no "twin paradox". Twins are aging in time-invariant space.

4. Conclusions

In today's physics we still think that with rope we measure distance in space and with clocks we measure distance in time. Einstein has kept time as the dimension of a four-dimensional continuum. My research confirms that this space-time continuum does not exist in the physical reality. Irreversible universal changes run in time-invariant universal space where black holes in the centre of galaxies are rejuvenating systems of the universe that is eternal and non-created.

References:

Einstein, A., Relativity: The Special and General Theory, Methuen & Co Ltd, p.93 (1916).

Fahr HJ, Heyl M. A universe with a constant expansion rate. Phys Astron Int J. 2020;4(4):156–163. doi:10.15406/paij.2020.04.00215

Fiscaletti D., Šorli A.S. **Searching for an adequate relation between time and entanglement.** *Quant. Stud. Math. Found.* **4, 357-374** (2017). doi: **10.1007/s40509-017-0110-5.**

Fiscaletti D. Sorli A.S., The Infinite History of NOW: A Timeless Background for Contemporary Physics, Nova Science Publishers (2014) ISBN: 978-1-63117-283-0

Harrison, E. Olbers' Paradox. *Nature* **204**, 271-272 (1964). **https://doi.org/10.1038/204271b0**.

Harari, Z. Analytical Resolution of the Dark Night Sky (Olbers') Paradox. *Astron. Nachr.* **340, 510-524 (2019).**
https://doi.org/10.1002/asna.201913540.

Knutsen, H. Darkness at night. EUR. J. PHYS. **18**, 295 (1997). **https://doi.org/10.1088/0143-0807/18/4/010**.

NASA:
https://wmap.gsfc.nasa.gov/universe/uni_shape.html (2014). Neil A., Relativity and GPS, Physics Today (2002). **https://doi.org/10.1063/1.1485583**.

Sarkar, S., and Jeffries, B. The solution to Olbers' paradox. PHYS. WORLD **15** (10), 17 (2002). **https://doi.org/10.1088/2058-7058/15/10/27.**

Sorli, A. S. (2020). Black Holes are Rejuvenating Systems of the Universe . *JOURNAL OF ADVANCES IN PHYSICS, 17*, 23–31.
https://doi.org/10.24297/jap.v17i.8620

Šorli, A., & Čelan, Štefan. (2020). The End of Space-time: Physics-Mathematics. *International Journal of*

Fundamental Physical Science, *10*(4), 31-34. https://doi.org/10.14331/ijfps.2020.330139

Sorli A., Fiscaletti D., Gregl T. New insights into Gödel's universe without time, Physics Essays, Volume 26, Number 1, March 2013, pp. 113-115(3), **https://doi.org/10.4006/0836-1398-26.1.113**.

Vidav I. (1959). Rešeni ni nerešeni problemi matematike (Solved and unsolved problems of mathematics), Mladinska knjiga

Time-invariant space and cosmology

Abstract

Results of several researchers suggest that time has no physical existence, that it is an illusion. Bijective research methodology confirms their results are right: time is what we measure with clocks; we measure with clocks the numerical sequential order of material change, i.e., motion running in time-invariant space. Time as the duration of change enters existence only when measured by the observer. The change runs only in time-invariant universal space. Humans are experiencing a run of changes in time-invariant space in the frame of the linear psychological time "past-present-future" that has its basis in the neurological activity of the brain. Time-invariant space is the fundamental arena of the universe. In the universe, there is neither a physical past nor physical future. The universe is what we can observe with our senses and measure with apparatuses. All the rest is pure speculation.

Keywords: space, time, change, cosmology, psychological time, bijectivity principle.

1. Introduction

Carlo Rovelli suggested that time is an illusion. On the other hand, he is not categorically denying the mainstream view that time cannot be eliminated from physics: "On the other hand, I also see well that the view I present here is far from being uncontroversial. Several authors maintain the idea that the notion of time is irreducible, and cannot be eliminated from fundamental physics" [1]. We will show in this article Rovelli is right about time being an illusion but still, time that we measure with clocks can remain in physics. The article introduces a model of time, which is an exact model of the time running in physical reality, using a *bijective research methodology* which is based on experimental results. There are no theoretical assumptions on time based on speculation, as for example what has been assumed for more than 100 years about time as the 4th dimension of space. There is no single experimental data confirming this last view.

Let's take the following example: a photon is moving in space from point A to point B. The distance *d* between A and B can be expressed as the sum of Planck lengths:

$$d = d_{p1} + d_{p2} + \cdots + d_{pn} = \sum_{i=1}^{n} d_{pi} \quad (1).$$

Photon is moving from d_{p1} to d_{p2} and so on; every Planck distance d_{pn} corresponds exactly to one Planck time t_P. In this sense, the Planck time is the fundamental unit of photon sequential motion from one to the next Planck distance. We can therefore write:

$$t = t_{p1} + t_{p2} + \cdots + t_{pn} = \sum_{i=1}^{n} t_{pi} \quad (2).$$

$$\frac{\sum_{i=1}^{n} d_{pi}}{\sum_{i=1}^{n} t_{pi}} = c \quad (3).$$

Photon is moving only in space and not in some physical time; the duration of photon motion from A to B in space is the sum of Planck times. Time is not continuous, is a discrete quantity; it is not some physical quantity that is running on its own, but the epiphenomenon of change, i.e. motion, its sequential numerical order. Time as the numerical sequential order of changes is a fundamental time; when fundamental time is measured by the observer, duration enters in existence. Duration is the so-called emergent time; there is no emergent time without the measurement of the fundamental time [2].

This article introduces a new research methodology, named *bijectivity principle*: let's consider the universe as a set X. In this set, we have four fundamental elements:

a) the energy in different forms: universal space is an energy structure, electromagnetic energy is a type of energy, dark energy is a type of energy, dark matter is a type of energy, energy in the form of matter;

b) the change;

c) the time;

d) the observer.

To build an adequate model of the universe in the set Y, which is the model of the universe, we must have there the same fundamental elements (Figure 1).

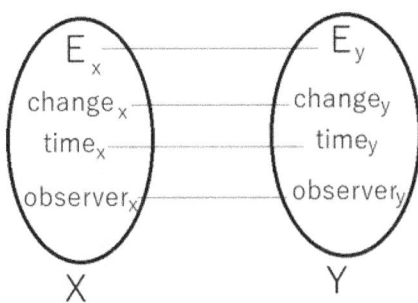

Figure 1: Bijective model of the universe.

$$X: \{E_x, C_x, t_x, O_x\} \quad (4a),$$
$$Y: \{E_y, C_y, t_y, O_y\} \quad (4b),$$

where E_x, C_x, t_x, O_x are energy, change, time and observer in the physical universe, and E_y, C_y, t_y, O_y the corresponding quantities in the model of the universe. We can write the following equation:

$$f(t_x) = t_y \quad (5).$$

The time in the model of the universe t_y is related to the time t_x in the physical universe by the bijective function of set theory. Bijectivity principle assures an adequate model of the universe; the physical universe and the corresponding model of the universe are related by a bijective function. In this bijective model of the universe, time in the physical universe has exactly the same meaning in the model of the universe: time is the sequential numerical order of material change, i.e., motion. Time is not continuous, it is discrete; Planck time is the fundamental unit of time.

The idea that space and time should be defined on the basis of set theory is not new. It was already proposed back in 1997 by Costa, Bueno, and French: "Science is the search for structure. Such a bold claim certainly deserves further elaboration and justification, but it is one which, it has recently been emphasised, has been held by a number of scientists and philosophers. Our aim in this note is two-fold. First, we wish to relate the consideration of such an aim to a long standing programme of work which has sought to develop a mathematically precise treatment of the notion of structure itself. In doing so, we shall analyse a particular illustrative case and thereby generate favourable evidence for the above claim as a whole. Thus, we shall consider the problem of constructing a set-theoretic structure, in Suppes's sense, which is capable of providing an axiomatic basis for those notions of space and time which underpin various theories of physics" [3]. We use in this research the bijectivity principle of set theory; the result is that universal space is a time-invariant. Time is the duration of material changes, i.e., motion in time-invariant space [4]. This model fits experimental physics and is in agreement with the mathematical formalism of Special Relativity and with the neuroscience research results on time.

Time as some physical quantity that is running in the universe was never observed experimentally and represents a "hard problem" of physics. In order to understand time, and to build an adequate model of it, we have to understand how we experience change, i.e., motion.

The human being perceives with senses the information about the change. The information is transformed into an electromagnetic signal and moves through nerves to the sight center in the brain. Here there is a neuronal activity that is continuously creating the psychological sensation of linear time: past, present, future. Linear time "past-present-future" is the result of neuronal activity of the brain. Several researches confirm that animal and human experience of linear time has the origin in neuronal activity of the brain [5,6]. We experience the material change, i.e., the motion in the frame of psychological time, which has its basis in neuronal activity; that is why we see changes running in some physical time despite there is no physical time in the universe (Figure 2).

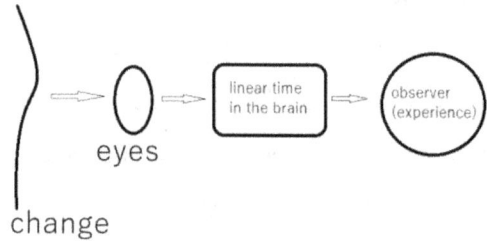

Figure 2: Linear time "past-present-future" is only in the brain.

Einstein was aware of the psychological time; he used to say: Past, present and future are stubbornly persistent illusions. Time has no independent existence apart from the order of events by which we measure it." [7]. In Minkowski manifold, the arena of Einstein's Special Relativity, the 4th dimension is not time, but is a spatial coordinate $X_4 = ict$, i.e., the product of the imaginary number i, the speed of light c and the time t as the numerical sequential order of the change. Minkowski manifold has not dimension $3D + t$; it is $4D$. In this four-dimensional continuum, time is the sequential numerical order of events.

Recent research confirms that by zapping the areas of visual cortex V5 with magnetic field, the duration of a given event that people observe cannot be well estimated. The duration of a given motion is defined by the neuronal activity of the visual cortex. When functioning of the visual cortex area is disturbed by the magnetic field, the experience of motion is disturbed too [8].

2. Time-invariant space impact on cosmology

The time-invariant space is the underlying reality of the universe. It has different names: "The Minkowski structure arises from deep reality itself, and is explored by physical entities with mass and photons of light. Human beings and more even human observers do not play any role in it. However, in the fact that we find the Minkowski metric so counter intuitive when we see it exposed, and more even when we see it drawn on a plane, the reason for this is deeply rooted in the nature of human observation. We are used to see paths in space only without involving the effect that velocity has on reaching into the future. The reason is that we are small, surrounded by all small physical entities, and we and all customary physical entities around us move 'in space' – remember that every physical entity moves with the velocity of light in time-space, or, more correctly in *the non temporal and non spatial underlying reality* – very very slowly only" [9], to the "*timeless configuration space*"[10], or "*superfluid quantum vacuum*" [11]. In this non temporal underlying reality, time travels are categorically excluded. One can travel only in this underlying reality and time is the duration of its motion. In this perspective Minkowski manifold is only a theoretical model that describes this underlying time-invariant reality. The formalism $X_4 = ict$ confirms that time t is the duration of photon motion. 4th dimension X_4 is not time ($X_4 \neq t$) and it is not "temporal", it is spatial in the same way as X_1, X_2, X_3 [12]. It was a historical mistake that Mikowski manifold was named "space-time". Minkowski manifold is four-dimensional geometry where time t is an element of the 4th dimension X_4.

The product of time t as the duration and light speed c is spatial distance d: $d = ct$. Einstein has added to this spatial distance imaginary number i, so we can write $X_4 = id$. This imaginary distance in the Minkowski model has no bijective correspondence with the physical world. Equation $f: id_X \to id_Y$ is false. Imaginary distance id cannot exist in the physical world.

The space-time interval S has been defined as: $S^2 = c^2 t^2 - (X^2 + Y^2 + Z^2)$. With the introduction of natural units: $c = \hbar = 1$, the space-time interval becomes: $S^2 = t^2 - (X^2 + Y^2 + Z^2)$ and the fourth coordinate of Minkowski manifold was interpreted as time t. The fourth coordinate of Minkowski manifold in its original form $X_4 = ict$ turned into $X_4 = t$ [4]. By taking natural units as 1, the time has been fully merged with space, and still today we think in physics that time is the 4th dimension of space-time that is the fundamental arena of the universe. For a century it was thought this is one of the most important achievements of physics. Today this achievement is questionable. We have to admit that rigorous bijective mathematical analysis confirms that time t in Minkowski manifold represents the duration of photon motion in time-invariant space; not more and not less. The idea that time is the 4th physical dimension of the universal space is one of the main obstacles to physics progress. Universal space is time-invariant, there is no physical past and there is no physical future. Material change run in time-invariant space. This result is a challenge for cosmology.

In the "block universe" model past and future are coexisting: "A new challenge emerged a century ago with Minkowski's space–time as it provided support for viewing the universe as a four-dimensional space–time block that exists as one entity. In this view, called the *block universe* (or *eternalism* in philosophical discussions), there is no basis for singling out a present time that separates the past from the future because all times coexist with equal status" [13]. In the bijective model of the "block universe" past and future are not coexisting in some eternal atemporal reality, they are non-existent. Irreversible universal change run in time-invariant space. When change $X + 1$ enters existence, the change X is not existing anymore. When change $X + 2$ is entering existence, the change $X + 1$ is not existing anymore. What really exists in the universe is what we can observe with our senses and measure with apparatuses. All the rest is pure imagination that is out of the frame of the rigorous bijective model of the universe. Symmetry in time that is represented by the formula $T: t \to -t$ is the model that has no bijective correspondence with the physical world [4]. In the universe, there is no symmetry in time because time is not a physical dimension in which the universe exists. Time is the duration of changes in time-invariant space of the "block universe".

Time-invariant space is replacing space-time as the fundamental arena of the universe. There is no such a thing in the universe as space-time and "slices of space-time".

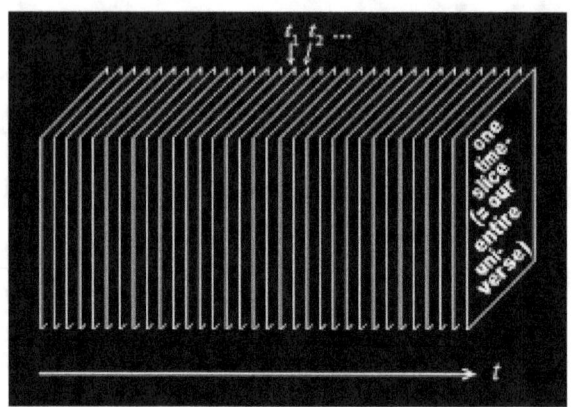

Figure 3: Hypothetical slices of space-time

There is no such thing in the universe as some physical "time distance" Δt between two events:

$$\Delta t = t_2 - t_1 \quad (6),$$

where t_1 is a given slice of space-time and t_2 is the next slice of space-time (see figure 3). The duration between two events Δt enters into existence in the act of the measurement from the side of the observer. All events in the universe happen in the same identical time-invariant space. Clocks tick in time-invariant space. Clocks do not measure some physical time running on their own. Duration enters existence when measured by the observer.

Figure 4: Time is the result of measurement with clocks

The universe is time-invariant. It does not run in some physical time; it runs in time-invariant space. Universal changes are irreversible, physical past is non-existent. Future also is non-existent. Time travel is out of the question. One can only travel in time-invariant space.

3. Entropy, gravity and entanglement in time-invariant space

In physics, we experience the increase of entropy and run of physical changes through the linear psychological time of "past-present-future"; we experience that changes are running in some linear time. We have seen in chapter three that the idea that entropy of a given system is increasing in time is the wrong imagination based on our experience of material changes run in the frame of psychological time. Considering that universal space is time-invariant, the universal changes run only in this time-invariant space and not in time. With clocks, we measure duration of events in time-invariant space. The arrow of time exists only in the form of mathematical arrow of the sequential numerical order of universal changes running in a time-invariant space. The entropy of a given system is increasing only in space and not in time; time is the numerical sequential order of entropy increasing [4].

The idea that time has no physical existence and that with clocks we measure internal relations between different physical changes running in a timeless space is entering in the mainstream of physics [14], so as the idea that gravity does not require time and is encoded in a timeless configuration of the universal space[15,16]. Time-invariant universal space is the direct information medium of entanglement by EPR-type phenomena [17]. For a century, entanglement has been difficult to understand because of a wrong image, i.e., that space-time (where time is a 4th physical dimension of space) is the fundamental arena of the universe.

It is shown in this article, universal changes are running in space that is time-invariant. Humans, we experience this time-invariance of universal space as "Now". Albert Einstein used to say: "that there is something essential about the Now which is just outside the realm of science" [18]. Neuroscience results that linear time "past-present-future" is created by the neuronal activity of the brain is bringing Einstein's Now into physics. Rovelli is right, time is an illusion in the sense, it has no physical existence. The mainstream did not understand well yet Rovelli's discovery [19]. Fiscaletti's research also confirms that in the universe there is no physical time [20, 21, 22]. What is important to understand is that considering physical time is an illusion, clocks are remaining useful tools for measuring the duration of changes in time-invariant space. That the duration is the result of the measurement is a fact we cannot ignore. This fact will deeply change our notion of the universe, life, and human being.

The study of the universe in some linear physical time past-present-future seems not correct because we do not have single experimental evidence that physical time exists. Universal changes run in a time-invariant universal space. Time (duration) enters existence only when measured by the observer. The universe is time-invariant. Thinking time start running after the initial explosion seems not appropriate. Time does not run in the universe; it runs only in the human brain.

4. Conclusions

In this article, it is shown that as suggested by Barbour, Gomes, Fiscaletti, and Rovelli there is no physical time out there in the physical reality. Changes run in time-invariant space that is the fundamental arena of the universe. This seems an important result for the physics and cosmology progress.

References:
1. Rovelli, C. "Forget time". *Found Phys* **41**, 1475 (2011). **https://doi.org/10.1007/s10701-011-9561-4**.

2. Fiscaletti, D., Sorli, A.S. Perspectives of the numerical order of material changes in timeless approaches in physics. *Found. Phys.* **45**(2), 105-33 (2015). **https://doi.org/10.1007/s10701-014-9840-y**.

3. da Costa, N.C.A., Bueno, O. & French, S. Suppes Predicates for Space-Time. *Synthese* 112, 271-279 (1997). **https://doi.org/10.1023/A:1004984927979**.

4. Šorli, A., & Čelan, Štefan. (2020). The End of Space-time: Physics-Mathematics. *International Journal of Fundamental Physical Science*, *10*(4), 31-34. **https://doi.org/10.14331/ijfps.2020.330139**

5. Buhusi, C., Meck, W. What makes us tick? Functional and neural mechanisms of interval timing. *Nat. Rev. Neurosci.* **6**, 755-765 (2005). **https://doi.org/10.1038/nrn1764.**

6. Mauk, M.D., Buonomano, D.V. The neural basis of temporal processing. *Annu. Rev. Neurosci.* **27**, 307-340 (2004). doi:10.1146/annurev.neuro.27.070203.144247.

7. Sorli, A.S., Klinar, D., Fiscaletti, D. New insights into the special theory of relativity. *Phys. Essays* **24**(2), 313-318 (2011). doi: 10.4006/1.3590161.

8. Protopapa, F., Hayashi, M.J., Kulashekhar, S., van der Zwaag, W., Battistella, G., Murray, M.M., Kanai, R., Bueti, D. Chronotopic maps in human supplementary motor area. *PLoS Biol.* **17**(3), e3000026 (2019). **https://doi.org/10.1371/journal.pbio.3000026.**

9. Diederik, A., Relativity Theory Refounded (2015). **https://arxiv.org/abs/1511.08735.**

10. Małkiewicz, P. Internal clocks in timeless universe. *J. Phys. Conf. Ser.* **880**, 012046 (2017). **https://iopscience.iop.org/article/10.1088/1742-6596/880/1/012046.**

11. Šorli, A.S. Mass-Energy Equivalence Extension onto a Superfluid Quantum Vacuum. *Sci. Rep.* **9**, 11737 (2019). **https://doi.org/10.1038/s41598-019-48018-2.**

12. Fiscaletti, D., Sorli, A. Bijective Epistemology and Space–Time. *Found Sci* **20,** 387–398 (2015). **https://doi.org/10.1007/s10699-014-9381-z.**

13. Vaccaro, J.A. The quantum theory of time, the block universe, and human experience. *Phil. Trans. R. Soc. A.* **376**, 20170316 (2018). **http://doi.org/10.1098/rsta.2017.0316.**

14. Barbour, J. The Nature of Time. arXiv:0903.3489v1 [gr-qc] (2009). **https://arxiv.org/abs/0903.3489**.

15. Gomes, H. Quantum gravity in timeless configuration space. *Classical Quant. Grav.* **34**(23) (2017). **https://iopscience.iop.org/article/10.1088/1361-6382/aa8cf9**.

16. Šorli, A.S. Mass–Energy Equivalence Extension onto a Superfluid Quantum Vacuum. *Sci Rep* **9**, 11737 (2019). **https://doi.org/10.1038/s41598-019-48018-2**.

17. Fiscaletti, D., Sorli, A.S. Searching for an adequate relation between time and entanglement. *Quantum Stud.: Math. Found.* **4**(4), 357-374 (2017). **https://link.springer.com/article/10.1007/s40509-017-0110-5**.

18. Smolin, L. *Time Reborn: From the Crisis in Physics to the Future of the Universe*, Houghton Mifflin Harcourt (2013), p. 93 ss.

19. Jaffe, A. The illusion of time. *NATURE* **556**, 304-305 (2018). **https://www.nature.com/articles/d41586-018-04558-7**.

20. **Fiscaletti, D.** Timeless Approach, The: Frontier Perspectives In 21st Century Physics, ISBN10 9814713155, World Scientific Publishing (2015).

21. Fiscaletti, D. Towards a Non-Local Timeless Quantum Cosmology for the Beyond Standard Model Physics, *Bulg, J. Phys. Vol.45 no.4 (2018), pp. 334-356*.

22. Fiscaletti, D. Sorli, A.S. The Infinite History of Now: A Timeless Background for Contemporary Physics, ISBN-10: 1631172832, Nova publishing (2014)

Advances of Relativity Theory

Abstract

Advances of Relativity Theory are in the replacement of the space-time model with time-invariant universal space that has a variable energy density. Every physical object with mass m and energy E is diminishing the energy density of space exactly for the amount of its energy. Lorentz factor has its origin in the variable density of universal space, we call it "superfluid quantum space" – SQS that is the primordial energy of the universe. Universal SQS is the absolute frame of reference for all observers as confirmed experimentally by the GPS system, which demonstrates that the relative rate of clocks is valid for all observers. A planet's perihelion precession and the Sagnac effect are the results of the SQS dragging effect.

1. Introduction

In Special Relativity, time t is the duration of photon motion along the 4th coordinate: $X_4 = ict$. This confirms that 4th coordinate is not time, 4th coordinate too is spatial distance. Duration t multiplied by light speed c is spatial distance X_4. Minkowski manifold X_1, X_2, X_3, X_4 has four spatial coordinates and thinking that time is its 4th coordinate is an utter mistake. Experimental data confirm that time t is a duration of motion in universal space that is time-invariant in the sense that time is not its 4th dimension. Development of Relativity Theory is based on three significant scientific discoveries:

- space-time is not a fundamental arena of the universe; time is not the 4th dimension of space. Time is merely the duration of motion in time-invariant universal space. Linear time "past-present-future" is phycological time based on the neuronal activity and exists only in the human brain. Irreversible universal changes run in time-invariant space. Time as the duration enters the existence when measured by the observer [1,2].
- Entanglement happens in time-invariant space only and not in time. Time-invariant universal space is the immediate medium of quantum entanglement [3].
- Universal space is not "empty", space is the fundamental energy of the universe, in today physics called "superfluid quantum vacuum" or "superfluid quantum space" [4]. We will call it in this article time-invariant superfluid quantum space – SQS.

GPS proves that the relative rate of clocks on satellites relative to the Earth's surface is valid for all observers, including observers in aeroplanes, trains, ships, and cars [5, 6]. This experimental fact, along with everyday experience, suggests a revision to our understanding of the famous *Gedankenexperiment* of one observer at a train station and another observer on a passing train. Standard physics textbooks describe that a clock at the station runs faster for the observer on the train, and the clock on the train runs slower for the observer at the station. In classic relativity both observers have their own 'internal time' inside the reference system in which they exist and both have an 'external time' that exists in the other observer reference system. This interpretation features four distinct times: the proper time of the observer at the station, the proper time of the observer in the train, the external time of the observer at the station, and the external time of the observer on the train. On the other hand, GPS proves that the relative velocities of clocks at the station and on the train are equally related to the rate of clocks on orbiting satellites, so are valid for both observers. If this were not so, then GPS could not work properly. In this article, we will develop a model where the relative rate of clocks in all inertial systems depends only on the variable energy density of superfluid dynamic space (SQS) and is valid for all observers.

2. Replacement of curvature of space with variable energy density of SQS

Einstein tensor has three elements, curvature tensor on the left, Einstein constant and stress-energy tensor on the right side of the equation:

$$G_{\mu\nu} = \kappa T_{\mu\nu} \quad (1).$$

Curvature tensor $G_{\mu\nu}$ describes curvature of space due to the presence of a given mass that is expressed by the stress-energy tensor $T_{\mu\nu}$. Curvature tensor is useful only on the macro scale, it cannot be applied on microscale, for example proton. In this article curvature tensor will be replaced by the minimal energy density of SQS in the centre of the given physical object with the rest mass m_0. This formula is valid from the proton to the supermassive black holes (SMBH). Every physical object with energy E and mass m is diminishing energy density of SQS in its centre exactly for the amount of its energy and correspondent mass:

$$E = mc^2 = (\rho_{Emax} - \rho_{Emin}) \cdot V \quad (2)\,[4],$$

$$\text{in units: } J = kg \cdot \frac{m^2}{s^2} = \left(\frac{J}{m^3}\right) \cdot m^3$$

where ρ_{Emax} is density of SQS in interstellar space, ρ_{Emin} is density of SQS in the centre of a given physical object and V is the volume of physical object.

SQS model distinguishes between rest mass and inertial mass. A given physical object with the rest mass m_0 is diminishing the energy density of SQS in its centre exactly for the amount of its energy E. The diminished energy density of SQS is creating the SQS pressure in the direction towards the centre of the physical object. This pressure is the common origin of the inertial mass m_i and of the gravitational mass m_g of a given physical object. Einstein has proved inertial mass and gravitational mass are equal, we confirm in this article they are equal because they have the same origin.

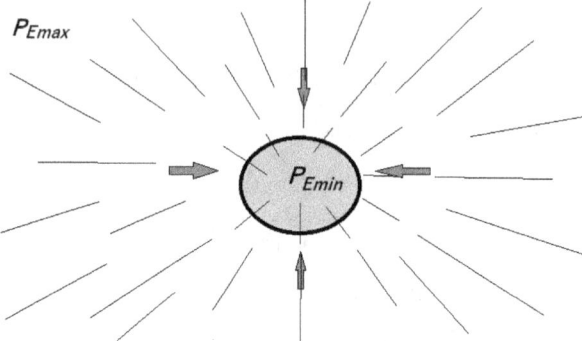

Figure 1: Inertial mass m_1 and gravitational mass m_g have the same origin in SQS pressure in the direction from p_{Emax} towards the direction p_{Emin}

We can calculate the energy density of SQS at a given point on the distance R from the centre of a given stellar object as follows:

$$\rho_R = \rho_{max} - \frac{3m}{4\pi(r+R)^3} \quad (3),$$

where m is mass of the stellar object, r is radius of the stellar object and R is the distance from the centre of the stellar object to the point where we calculate ρ_R density of SQS (Figure 2).

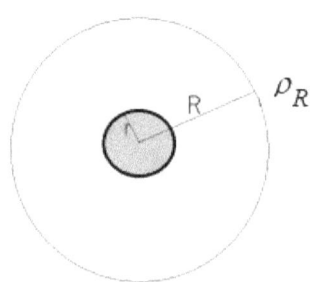

Figure 2: Energy density of SQS at the distance R from the centre.

In Advanced Relativity curvature of space is replaced with variable energy density of space. More space is curved in GR, less is its energy density in Advanced Relativity. When R is zero, we have minimal energy density in the centre of a given stellar object and maximum curvature of space in GR. When R is equal to the radius r of the stellar object, we have energy density of space on the surface of the stellar object. When R is close to infinity, we have the maximum energy density of superfluid quantum space (SQS) in interstellar space where the curvature of space in GR is at the minimum.

3. The Lorentz factor and variable energy density of the SQS

Lorentz factor γ expresses a diminished rate of clocks and a diminished velocity of material changes due to the motion. In the famous example of a train passing a station, t' is the elapsed time on the train and t is the elapsed time at the station, such that:

$$t' = \frac{1}{\sqrt{1-\frac{v^2}{c^2}}}\left(t - \frac{vx}{c^2}\right) \quad (4),$$

where $\frac{1}{\sqrt{1-\frac{v^2}{c^2}}}$ is the Lorentz factor γ, v is velocity of the train and x is the distance along the motion from the station clock to the clock in the train [7]. This diminished rate of clocks on the train has its origin in the decreased energy density of the SQS inside the train. In general, a moving system interacts with the SQS energy so that the higher the velocity v, the stronger the interaction and more SQS energy is integrated into the moving object, which in turn increases its mass m of a moving object according to:

$$m = \gamma m_0 = m_0 + \frac{EK}{c^2} \qquad (5),$$

where m_0 is the object's rest mass, EK is moving object kinetic energy in the form of integrated energy of SQS and γ is the Lorentz factor.

Out of equation (2) follows, the equation for the minimal energy density of SQS ρ_{Emin} in the rest wagon of the train is following:

$$\rho_{Emin} = \rho_{Emax} - \frac{m_0 c^2}{V} \qquad (6).$$

Formula for the energy density of SQS in the moving wagon $\rho_{Emin.m}$ is following:

$$\rho_{Emin.m} = \rho_{Emax} - \gamma \frac{m_0 c^2}{V} \qquad (7),$$

where $\rho_{Emin.m}$ is the additionally diminished energy density in the centre of the wagon because moving wagon matter is absorbing some of the SQS energy which increases wagon's relativistic mass accordingly go the Eq. (5). This decreased energy density of the SQS $\rho_{Emin.m}$ causes the rate of the clock on the moving wagon to run slower. According to Equation (7), we can express the Lorentz factor as follows:

$$\gamma = \frac{(\rho_{Emax} - \rho_{Emin.m})V}{m_0 c^2} \qquad (8).$$

The difference in the energy density of SQS we can write as $(\rho_{Emax} - \rho_{Emin.m}) = \Delta\rho_E$.

By replacing the $m_0 c^2$ with energy E of the rest object we get:

$$\gamma = \frac{\Delta\rho_E V}{E} \qquad (9),$$

where E is the energy of the object at rest, V is the volume of the object and $\Delta\rho_E$ is the difference between the energy density of SQS far away from the physical object and the centre of the moving object. In Eq. (9) rest energy E and volume V of the object are not changing. The only parameter that changes the Lorentz factor is the diminished energy density of SQS in the centre of the moving object which depends on the velocity v of the object. So, the higher is the speed v, the stronger is the interaction of the object with the SQS, absorption of the SQS energy is greater, and the energy density of the SQS in the centre of the moving object becomes smaller. With a smaller density of the SQS in the centre of the wagon (and in any other moving object), the rate of the clock is slower:

increased velocity → increased absorption of the SQS energy
→decreased energy density of the SQS→ decreased rate of a clock.

In the SQS model, the relative rate of clocks and the relative velocity of material changes depend on the variable energy density of the SQS. For example, muon decay when approaching the Earth's surface decreases because their velocity increases [8]. With this increasing velocity, the minimal energy density of the SQS of the muon decreases, and the time of decay increases. The duration of this decay does not depend on the selected reference system nor the chosen observer. A muon's relativistic decay is valid for all observers and is determined only by the variable energy density of the SQS.

4. Advances of Special Relativity

In the areas of universal space where energy density of SQS is not changing the speed of light is constant for all observers because all observers exist in the same SQS and light is the vibration of the SQS. The velocity of light in the intergalactic space is constant, the energy density of SQS there is at the maximum. In the areas where the energy density of SQS is lower gravity is stronger and light speed diminishes minimally. We call this effect in classic relativity wrongly "gravitational time dilation", see chapter 5.2; what Shapiro has measured is that in stronger gravity light needs more time to travel on a given distance which means that its speed has a minimal diminishment.

The area of SQS around a given physical object is moving and rotating with it. We call this "dragging effect". SQS around the Earth is rotating and so the light motion needs a shorter duration when travels in the direction of Earth's motion because SQS is also rotating with the Earth. When light is moving in the opposite direction of Earth motion from B to A needs a longer duration. In both cases light speed is constant. By assuming constancy of light in stationary SQS and in moving SQS, we will develop an SR theory without contradictions as those that exist with the current SR in the thought experiment of two-photon clocks.

Here, we place two identical photon clocks on a moving train where one is positioned horizontally in the direction of motion, and the other is positioned vertically. According to the idea of "length contraction," the horizontal photon clock will shorten in length and tick faster compared to the vertically oriented clock that will not diminish in length. This scenario leads to a contradiction as SR does not predict that the two clocks in the same inertial system will have different rates. The solution is available through the development of an SR model in a three-dimensional Euclidean space with a Galilean transformation and Selleri's equation for the variable rate of clocks with no occurrence of "length contraction": "Einstein's formalism of special relativity based on the standard Lorentz transformations may be derived from a more fundamental 3D Euclidean space, with Galilean transformations for the three spatial dimensions and Selleri's transformation for the rate of clocks" [9]:

$$t' = \sqrt{1 - \frac{v^2}{c^2}}\, t \quad (10).$$

Selleri's equation (10) confirms that rate of clocks is not related to the spatial dimensions. SR equipped with this formalism also describes successfully all phenomena in 3D Euclidean space previously described by classical SR such as aberration of light, Doppler effect, Jupiter's satellites occultation and radar ranging of the planets [9]. By the use of algebra equation (10) can be derived from equation (4) [7].

A second contradiction occurs with the rate of the vertical photon clock on the moving train from the perspective of the observer at the station. The classical interpretation states that for the observer at the station, the vertical photon clock ticks slower because they see the photon in the clock moves in a 'zig-zag' direction [10], as illustrated in Figure 3.

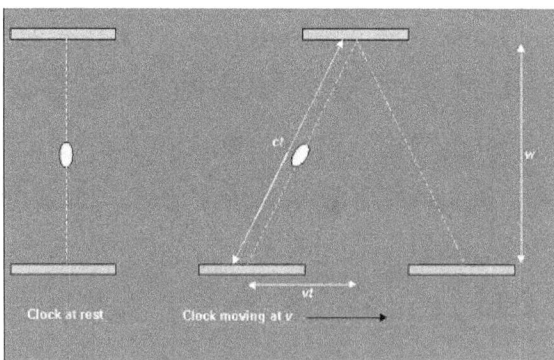

Figure 3. An observer at rest, seeing a moving clock photon.

This explanation may appear illogical because the optical illusion of the stationary observer cannot slow the rate of the moving clock. Instead, in the moving train, the energy density of the SQS diminishes, causing a reduced velocity of the photon. With the diminishing of SQS energy density also the velocity of light diminishes a bit (see chapter 5.2.). Therefore, the moving vertical photon clock ticks slower in the moving train because of the diminished energy density of SQS and not because of the optical illusion of the stationary observer in the station.

In Advanced Relativity rate of clocks in all moving inertial systems depends on the diminished energy density of SQS in the system due to its motion. The relative rate of clocks does not depend on the position of a given observer and is valid for all observers. GPS is proving this without any doubt. Because of the GR effect clocks on the satellites are running faster than clocks on the Earth surface for 45 microseconds per day. Because of the SR effect clocks are running slower on the satellites than the clocks on the Earth surface for 7 microseconds per day [5]. This is valid for all observes.

If the clock would be taken out of the satellite it would keep the same rate. The mass of the satellite is too small to influence the rate of the clock because of the GR effect and SR effect. The only factor that determines diminished SQS energy density and so the Lorentz factor γ and consequently the rate of clock related to the SR effect is the velocity v of the clock. We have shown in our article the relation between the Lorentz factor and diminished energy density of SQS in equation (9).

In Advanced Relativity "length contraction" and "time dilation" where time is supposed to be the 4th dimension of space are abolished. We do not know a physical mechanism that would shorten the length of the objects that are moving in the direction of motion. The idea was created by Hendrik Lorentz in 1892 to save "ether". After Michelson-Morley's experiment has given a null result, Lorentz predicted that the beam in the interferometer that was pointed in the direction of Earth motion has shortened. In Advanced Relativity time is the duration of material change, i.e., motion in time-invariant SQS and cannot dilate. Time as duration is the result of the measurement form the side of the observer and as such has no physical existence [1,2]. What is "relative" in the universe is the velocity of material changes that depends on the variable energy density of SQS.

5. Advances of General Relativity

In the model presented in this paper, the rotation of stellar objects also causes rotation of the surrounding SQS. For example, the rotation of the SQS around the Sun causes precession of the planets according to the following equation:

$$\sigma = \frac{24\pi^3 L^2}{Tc^2(1-e^2)} \quad (11),$$

where the perihelion shift σ is expressed in radians per revolution, L is the **semi-major axis**, T is the **orbital period**, c is the speed of light, and e is the **orbital eccentricity** [11]. The mass of the Sun is not included as there is also no mass of a planet, so these masses do not affect the precession of the planets. In the model of SQS, the perihelion shift σ depends on the rotation of the SQS caused by the rotation of the Sun, which in turn pushing the planets and causes a perihelion precession. With increasing distance from the Sun, the impact of the rotating SQS on planets diminishes along with the precession of the perihelion.

Irregular and spiral galaxies comprise approximately 60% of all galaxies in the universe. In the centre of most spiral galaxies exist a rotating black hole [12]. We suggest in this article that rotating black holes are rotating the surrounding SQS. This might be one of the physical causes of their spiral shape; dragging effects between the rotating black hole and rotating SQS diminishes with the distance from the black hole leading to the spiral geometry. The development of the mathematical model of this effect is one of the goals of our further research.

In 2019 NASA has reported: "as if black holes weren't mysterious enough, astronomers using NASA's Hubble SQS Telescope have found an unexpected thin disk of material furiously whirling around a supermassive black hole at the heart of the magnificent spiral galaxy NGC 3147, located 130 million light-years away. The conundrum is that the disk shouldn't be there, based on current astronomical theories" [13]. Our proposal to solve this conundrum is that in current astronomical theories, supermassive black holes rotate in an "empty space." In SQS model presented in this article, supermassive black holes rotate in the medium of SQS, and their rotation, in turn, rotates the SQS. Therefore, this dragging effect of SQS might be a physical cause of thin disc that is furiously whirling around a supermassive black hole of the spiral galaxy NGC 3147.

In GPS, Sagnac effect corrections make the system work [14]. Essentially, a signal when moving from A to B in the direction of Earth's rotation needs less time compared to when moving from B to A in the direction opposing Earth's rotation. In a SQS model presented in this article, light has a constant speed regardless of the SQS's motion. So, when moving from A to B, light needs less duration (time) because it is moving in the same direction as the SQS.

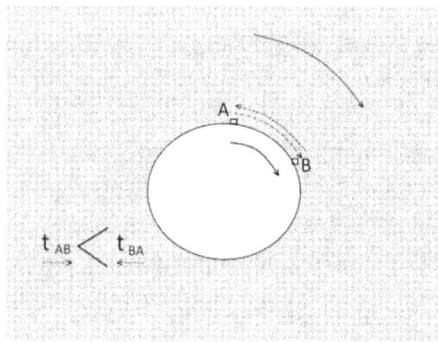

Figure 4. A light signal's duration due to the rotation of the quantum SQS.

Sagnac's experiment with the rotating interferometer is indisputable proof that photon does not move in the empty space deprived of physical properties [15,16]. On the contrary, it proves that the photon is the excitation of the SQS that is dragged by the rotating interferometer.

The Michelson-Morley experiment demonstrated a null result because the area of the SQS around the Earth is not only rotating with the Earth but is also moving with the Earth, as shown in Figure 5.

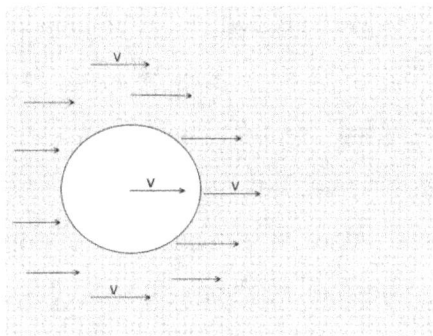

Figure 5. The SQS moves with the Earth.

So, the negative outcome of Michelson-Morley abolished the ether model. According to bijective research methodology, where every element in the model has exactly one correspondent element in physical reality, universal space is neither filled with ether, nor is it 'empty.' Instead, universal space contains material objects that contain energy. Energy and matter cannot exist in an 'empty' space deprived of all physical properties, so in this article universal space, we name it SQS, is understood as the primordial energy of the universe. A photon is a wave of the SQS, and the velocity of this photon wave is the speed of light, c. The photon velocity c is invariant with respect to the SQS's motion as it appears through the Sagnac effect. The photon velocity diminishes minimally when a photon moves through a stronger gravity where the density of the SQS is lower, as is the case with the Shapiro experiment, as will be described in Section 5.2.

Motion and rotation of the universal space with physical objects is referred to as the 'SQS dragging effect' in this article. Dragging effect was measured by Josef Lense and Hans Thirring in 1918 and was called 'frame-dragging' due to the belief that space-time being distorted by rotating objects, reference [17]. Recent research confirms that this 'space-time' model has no physical reality [3], so it cannot be dragged by rotating or moving objects. According to bijective research methodology, an adequate term would be the 'SQS dragging effect.'

In this article, the idea of "length contraction" and "time dilation" do not exist. Length contraction in SR is only a mathematical tool with no physical reality. By using "length contraction," Einstein achieved a constancy of light in all inertial systems. On the other hand, all observers measure the same value for the velocity of light because the light is the vibration of the SQS in which all observers move. Also, time, being in Special Relativity the fourth dimension of space, does not "dilate"; relative velocity of material changes (the rate of clocks included) depends on the variable density of the SQS.

5.1. Gravitational redshift

Gravity has its origin in the variable energy density of SQS. SQS fluctuations are directed from the higher energy density of SQS towards the lover density of SQS. Therese fluctuations that interact with photons to diminish their frequency, which is referred to as 'gravitational redshift.' When light from distant galaxies reaches the Earth, its frequency is lower. On its path to Earth, light loses some of its energy because it is moving against the SQS fluctuations that points toward the direction of galaxies, so that

$$E_{photon\,Earth} = E_{photon\,galaxy} - \Delta E \qquad (12),$$

where $E_{photon.galaxy}$ is the energy of the photon at the galaxy, $E_{photon.Earth}$ is the energy of the arrived photon at the Earth, and ΔE is the loss of energy due to the fluctuations of the SQS,

$$\Delta E = h\Delta v \qquad (13),$$

where h is Planck's constant and Δv is the decrease of the photon frequency due to SQS fluctuations.

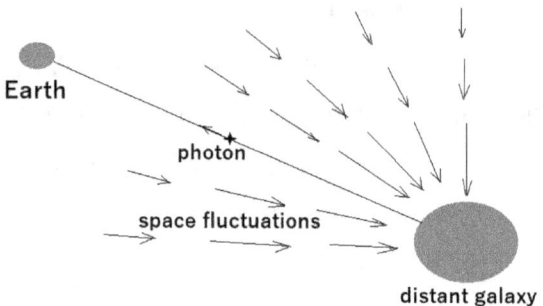

Figure 6. The redshift of light arriving from galaxies caused by SQS fluctuations.

Because of different densities of the SQS, the frequency of light also changes when moving from the source to the receiver above the Earth's surface. In a Harvard University experiment, a source on the Earth's surface and a receiver at the height of 22,5 meters were positioned, as illustrated in Figure 5.

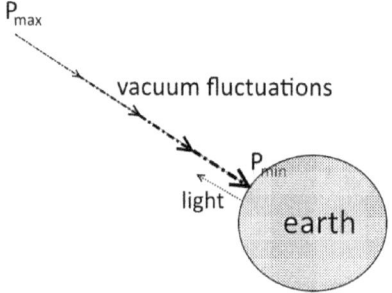

Figure 7. The redshift of light moving from the Earth's surface upwards.

The Mössbauer effect was used to measure the difference between γ-ray emission and absorption frequencies at each end of the experiment. The measurement accuracy was $\Delta\omega/\omega \approx 10^{-15}$, which shows a change of light frequency as

$$\frac{\Delta\omega}{\omega} = \frac{GM}{R^2 c^2} h \quad (14),$$

where M and R are the mass and radius of the Earth, respectively [18].

We can substitute into Equation (14) for the Earth mass M with the $\frac{(\rho_{Emax} - \rho_{Emin})V}{c^2}$ from Equation (2) as:

$$\frac{\Delta\omega}{\omega} = \frac{G(\rho_{Emax} - \rho_{Emin})V}{R^2 c^4} h \quad (15),$$

which can be expressed as:

$$\frac{\Delta\omega}{\omega} = \frac{G(\rho_{Emax} - \rho_{Emin})4\pi R^3}{3R^2 c^4} h$$

$$\frac{\Delta\omega}{\omega} = \frac{4\pi RG(\rho_{Emax} - \rho_{Emin})}{3c^4} h \quad (16).$$

Equation (16) confirms that gravitational redshift depends on the minimal energy density of the SQS ρ_{Emin} in the Earth's centre. SQS fluctuation in the direction from ρ_{Emax} towards ρ_{Emin} are explaining so called "tired light" model of astronomer Fritz Zwicky. Zwicky proposed that light is losing some of the frequency when travelling vast distances from the galaxies to the planet Earth [19].

In model of Relativity here presented SQS fluctuations from ρ_{Emax} towards ρ_{Emin} are causing the Pioneer anomaly which means the observed deviation from predicted accelerations of the Pioneer 10 and Pioneer 11 spacecraft after they passed on their trajectories out of the Solar System [20]. SQS fluctuations represent the barrier for the photons and are also representing the barrier for the Pioneer spacecraft slowing down their acceleration.

5.2. Shapiro gravitational delay

In 1964, Shapiro [21] measured the decreased velocity of light in a gravitational field, as observed by the speed of a light signal diminishing when passing the gravitational field of the Sun. Shapiro's result is understood by today's physics as a 'gravitational time delay' caused by spacetime dilation, which increases the path length. According to bijective research methodology, where every element in the model have the exact correspondent element in physical reality, this interpretation appears not to be exact as Shapiro did not measure spacetime dilation. In SR, the element of 'spacetime dilation' has no bijective correspondence in the physical world as it has never been observed in physics that spacetime or space are dilating. According to bijective research methodology, Shapiro's result should be termed the 'gravitational diminishing of light-speed' caused by the diminished energy density of the SQS. In SQS with a given gravitational field, the energy density of the SQS is diminished that causes a minimal diminishing of the speed of light as defined by the permittivity and permeability of the SQS:

$$c = \frac{1}{\sqrt{\mu_0 \varepsilon_0}} \quad (17),$$

where μ_0 is the magnetic permeability and ε_0 is electric permittivity of the SQS where there is no influence of gravity, and the density of the SQS is at its maximum ρ_{Emax}. In the SQS with a gravity field, the energy density of the SQS decreases and causes minimal diminishing of the permittivity and permeability, which in turn result in the minimal diminishing of the speed of light, as presented by Masanori, "it is known that the speed of light depends on the gravitational potential. In the gravitational fields, the speed of light becomes slow, and time dilation occurs. In this discussion, the permittivity and permeability of free space are assumed to depend on gravity and are variable" [22]. Minimal variability of the speed of light caused by a gravity field maintains SR because its first postulate is valid only in space where gravity is absent. The electric permittivity in flat space with no gravity is ε_0, and magnetic permeability in flat with no gravity SQS is μ_0. Following Puthoff, on the surface of stellar object, permittivity and permeability are:

$$\varepsilon = K\varepsilon_0 \quad (18),$$
$$\mu = K\mu_0 \quad (19),$$

where the space dielectric constant K on the surface of a stellar object is:

$$K \approx 1 + \frac{2Gm}{rc^2} \quad (20),$$

with G being the gravitational constant, M is the mass, and r the distance from the origin located at the centre of the mass M [18]. Combining equation (2) and (20) we can write following equation:

$$K \approx 1 + \frac{2G\,(\rho_{Emax} - \rho_{Emin})V}{rc^4}$$

which shows the SQS dielectric constant depends on the variable energy density of SQS. In this sense, a diminished energy density of the SQS on the surface of a given stellar object increases permittivity and permeability of the SQS which, in turn, minimally decreases the velocity of light as:

$$c = \frac{1}{\sqrt{\varepsilon \mu}} \qquad (22),$$

where ε is the electric permittivity and μ is its magnetic permeability of the SQS where there is gravitational field. From this, it follows that the Shapiro gravitational time dilation has its origin in the diminished energy density of the SQS near the stellar objects, which increases the dielectric constant K of the SQS and this minimally decreases the velocity of light. In other words: diminishing SQS energy density → increases the dielectric constant → increases the electric permittivity of the SQS → increases the magnetic permeability of the SQS → decreases the velocity of light.

The classic textbook explanation of the Shapiro experiment is that in stronger gravity, time, as the fourth physical dimension of space, dilates causing light to need more time to reach the point B from point A in a space-time that acts as the fundamental arena of the universe. This article shows that through a bijective interpretation of data, where data are not interpreted but read directly, requires an exact explanation where the velocity of light is minimally diminishing in a gravity field due to a diminished energy density of the SQS.

Doppler effect proves the second postulate of SR, which states that "the speed of light c is a constant, independent of the relative motion of the source." The observer exists in SQS, and a photon is the vibration of the same SQS. When the observer moves toward or away from the source of light, they will experience the Doppler effect. With the understanding that the moving observer and the source both exist in the same SQS and that light is the vibration of the SQS, the second SR postulate becomes logical. The observer sees the light with a given frequency coming from the source. When the observer moves away from or closer to the source, the frequency of the light diminishes or increases, respectively.

5.3. Gravitational lens

SQS fluctuations bend light, which we refer to as a 'gravitational lens,' and this bending of light as it passes the Sun is one proof of General Relativity. The SQS fluctuations near the Sun's surface are strongest and push the photons, causing them to bend, as illustrated in Figure 8.

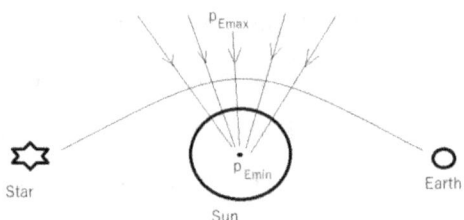

Figure 8. SQS fluctuations bending light around the Sun.

SQS fluctuations bend the photon's trajectory, which we call 'gravitational lens.' Einstein's formula for the bending of light as it passes the Sun [23] is expressed as:

$$\delta = \frac{4G M_s}{c^2 b} \qquad (23),$$

where δ is the angle of deflection, M_S is the mass of the Sun, c is the speed of light, and b is the minimum distance between the trajectory and the centre of the Sun. The mass of the Sun, M_S, can be expressed according to Equation (2), which we can combine with Equations (23) and to obtain

$$\delta = \frac{4G\,(\rho_{Eman} - \rho_{Emin})V}{c^4 b} \qquad (24).$$

Equation (24) confirms that SQS fluctuations that carry gravity are directed from SQS where the energy density of the SQS is ρ_{Emax}, towards the energy density of the SQS is ρ_{Emin} in the centre of the Sun, as in Figure 8. These SQS fluctuations push the photons, causing light deflection. Light passing the Sun is not deflected as a result of the curvature of universal space; measurements by NASA have proven that the universe's space has a Euclidean shape [24]. Light is deflected around gravitational objects, such as the Sun, due to a push from SQS fluctuations.

5.4. Gravitational waves are waves of superfluid quantum space

In this article gravitational waves are not represented as "ripples in the fabric of space-time" [25] because space-time, as the fundamental arena of the universe, does not exist. Gravitational waves are ripples of the SQS. Gravitational waves change the permittivity and permeability of SQS. As gravitational waves enter the LIGO interferometer, they changed permeability and permittivity of the SQS which minimally changes the speed of light moving in the beams of the interferometer. This minimal change in the speed of light is what is directly measured by LIGO. No direct data exists to confirm that the length of the beams of the interferometer change due to the gravitational waves. How the subtle phenomena of a gravitational wave could shrink or elongate the length of the interferometer beams, which have a solid iron-concrete core, is an unanswered question. The model here presented solves this question through the direct reading of the available data. What is measured by

LIGO is the minimal change in light speed due to minimal variations of the permittivity and permeability of the SQS caused by the gravitational wave entering the interferometer.

Our research suggests photons are excitations of superfluid quantum space (also named "superfluid quantum vacuum") [26]. Recent research confirms gravitational waves have a speed close to the speed of the photon, "since the recent major discovery in physics, the first measurement of gravitational waves, achieved by the LIGO/Virgo collaboration, several events have been registered. In particular, the merging of two neutron stars detected with its electromagnetic counterpart by the FERMI satellite has led to implications of paramount importance. One of them is the speed of gravitational waves now constrained to be extremely close to that of light, at the 10−15 level, at low redshifts" [27]. In the model presented in this article the photon and gravitational wave are both excitations of the SQS.

6. Conclusions

Classical Theory of Relativity is a primarily mathematical theory with several elements that have no physical existence as for example "time dilation", "length contraction", "gravitational time dilation", "curvature of space". In Advanced Relativity, these elements are removed from the model. Time dilation and gravitational time dilation are replaced with a variable velocity of material changes (light speed included) that depend on the variable energy density of SQS. The curvature of space is replaced by the variable energy density of SQS that is valid from the scale of the proton to the scale of the black hole. This is the new perspective to unify quantum mechanics and relativity theory.

References:
1. Fiscaletti, D., Sorli, A. Perspectives of the Numerical Order of Material Changes in Timeless Approaches in Physics. *Found Phys* **45,** 105–133 (2015). **https://doi.org/10.1007/s10701-014-9840-y.**
2. Šorli, A., & Čelan, Štefan. The End of Space-time: Physics-Mathematics. *International Journal of Fundamental Physical Science, 10*(4), 31-34. (2020). **https://doi.org/10.14331/ijfps.2020.330139.**
3. Fiscaletti, D., Sorli, A. Searching for an adequate relation between time and entanglement. *Quantum Stud.: Math. Found.* **4,** 357–374 (2017). **https://doi.org/10.1007/s40509-017-0110-5.**
4. Fiscaletti, D., & Sorli, A. Quantum Relativity: Variable Energy Density of Quantum Vacuum as the Origin of Mass, Gravity and the Quantum Behaviour. *Ukrainian Journal of Physics, 63*(7), 623. (2018). **https://doi.org/10.15407/ujpe63.7.623**
5. Neil A., Relativity and GPS, Physics Today (2002). **https://doi.org/10.1063/1.1485583.**
6. Agnew, D.C., Larson, K.M. Finding the repeat times of the GPS constellation. *GPS Solut* **11,** 71–76 (2007). **https://doi.org/10.1007/s10291-006-0038-4.**
7. Sorli, A., Klinar, D., Fiscaletti, D., New insights into the special theory of relativity, Phys. Essays **24,**2 (2011). **https://doi.org/10.4006/1.3590161.**

8. Liu L., The Speed and Lifetime of Cosmic Ray Muons, (2007) http://web.mit.edu/lululiu/Public/pixx/not-pixx/muons.pdf.
9. Fiscaletti D., Sorli A., About a new suggested interpretation of special theory of relativity within a three-dimensional Euclid space. Annales Universitatis Mariae Curie-Sklodowska, sectio AAA–Physica, 68: 39-62. (2015). https://journals.umcs.pl/aaa/article/view/806/649.
10. Philip Harris, Special Relativity, pp. 29 https://web.stanford.edu/~oas/SI/SRGR/notes/srHarris.pdf
11. Einstein, Albert (25 de noviembre de 1915). «Die Feldgleichungun der Gravitation». Sitzungsberichte der Preussischen Akademie der Wissenschaften zu Berlin: 844-847 (1915).
12. Loveday J., The APM Bright Galaxy Catalogue, **Monthly Notices of the Royal Astronomical Society. 278** (4): 1025–1048. **arXiv:astro-ph/9603040** (1996).
13. NASA, Hubble Uncovers Black Hole Disk that Shouldn't Exist, (2019). https://www.nasa.gov/feature/goddard/2019/hubble-uncovers-black-hole-that-shouldnt-exist
14. Ashby N., The Sagnac Effect in the Global Positioning System. In: Rizzi G., Ruggiero M.L. (eds) Relativity in Rotating Frames. Fundamental Theories of Physics, vol 135. Springer, Dordrecht (2004). https://doi.org/10.1007/978-94-017-0528-8_3.
15. Sagnac, G., L'éther lumineux démontré par l'effet du vent relatif d'éther dans un interféromètre en

rotation uniforme, in *C.R. Acad. Sci.* (Paris) 1913, t. 157, pp. 708–710.
16. Sagnac, G., Effect tourbillonnaire optique. La circulation de l'éther lumineux dans un interférographe tournant, *J. Phys. Radium*, 1914, Ser. 5, t. 4, pp. 177–195.
17. *Lense, J., Thirring H., "Über den Einfluss der Eigenrotation der Zentralkörper auf die Bewegung der Planeten und Monde nach der Einsteinschen Gravitationstheorie". Physikalische Zeitschrift. **19**: 156–163.* [On the Influence of the Proper Rotation of Central Bodies on the Motions of Planets and Moons According to Einstein's Theory of Gravitation] *(1918).*
18. Puthoff H.E., Polarizable-SQS(PZ) presentation of general relativity, Found.Phys. 32, 927-943. (2002). **10.1023/A:1016011413407**.
19. Zwicky F., On the Red Shift of Spectral Lines through Interstellar Space. PNAS **15**:773–779. (1929). **10.1073/pnas.15.10.773**.
20. *Nieto, M. M.; Turyshev, S. G., Finding the Origin of the Pioneer Anomaly.* **Classical and Quantum Gravity**. *21(17): 4005–4024. (2004).* https://iopscience.iop.org/article/10.1088/0264-9381/21/17/001.
21. *Shapiro I.I., Fourth Test of General Relativity.* **Physical Review Letters**. *13 (26): 789–791. (1964).* https://doi.org/10.1103/PhysRevLett.13.789.
22. Masanori S., Gravitational effect on the refractive index: A hypothesis that the permittivity, ε0, and permeability, μ0 are dragged and modified by the gravity **https://arxiv.org/vc/arxiv/papers/0704/0704.1942v3.pdf**

23. Albert Einstein, Lens-like action of the star by the derivation of light in the gravitational field, *Science*, Vol. 84, Issue 2188, pp. 506-507. (1936).
24. NASA, (2014). **https://wmap.gsfc.nasa.gov/universe/uni_shape.html**
25. LIGO, (2008). Gravitational Waves: Ripples in the fabric of SQS-time, **https://SQS.mit.edu/LIGO/more.html**
26. Fiscaletti, D., & Sorli, A. A Three-Dimensional Non-Local Quantum Vacuum as the Origin of Photons. *Ukrainian Journal of Physics*, *65*(2), 106. (2020). **https://doi.org/10.15407/ujpe65.2.106**.
27. **Louis Perenon, Julien Bel, Roy Maartens, Alvaro de la Cruz-Dombriz**, Optimising growth of structure constraints on modified gravity, (2019). **https://arxiv.org/abs/1901.11063**

Time-invariant Superfluid Quantum Space as the Unified Field Theory

Abstract

The novelty of 21st-century physics is the development of the "superfluid quantum vacuum" model, also named "superfluid quantum space" that is replacing space-time as the fundamental arena of the universe. It also represents the model that has the potential of unifying four fundamental forces of the universe. Superfluid quantum space is represented as the time-invariant fundamental field of the universe where time is merely the duration of material changes.

Keywords: unified field model; superfluid quantum space; gravity; entanglement; time.

1. Introduction

Valeriy Sbitnev suggests that superfluid quantum vacuum also named superfluid quantum space is the physical origin of the universal space [1,2]. In this article we developed a model of the time-invariant n-dimensional complex superfluid quantum space which offers the new solution for Einstein's dream of a "Unified Field Model". In Einstein's Relativity the universal space is understood as a 4-D reality with tree spatial dimensions and one temporal dimension. Bezuglav also suggested that the superfluid quantum vacuum, which is the physical origin of the universal space, is four-dimensional [3]. In experimental physics, time is duration of material change, i.e., motion in space. Taking this in account we developed the model of time-invariant n-dimensional complex superfluid quantum space, shortly "SQS".

The measured value of cosmological constant $\Lambda = 5.96 \cdot 10^{-27}$ kg/m³ [4] is different from its calculated value following the Planck metrics for the magnitude of 10^{123}; this discrepancy is an unsolved subject of physics for decades [5]. Regarding the suggested energy density of space proposed in this article, we are defending our proposal by the fact that the gravitational constant G is obtained by measurement and is expressed by the Planck energy density ρ_{EP} and the Planck time t_P as [6]:

$$G = \frac{c^2}{\rho_{EP} t_P^2} \quad (1).$$

This means that the Planck energy density ρ_{EP} reflects the real energy density of a 4-D universal space. In the absence of stellar objects, the energy density of the universal space has a value of Planck energy density which is $\rho_{EP} = 4.64 \cdot 10^{113} Jm^{-3}$ [6]. Meis has developed another formula for calculating the gravitational constant G:

$$G = \frac{l_P^2 c^2}{4 \pi e \xi} \quad (2)$$

where e is the elementary charge constant and ξ is the vector potential amplitude of the electromagnetic field to a single photon state ($\xi = 1.747 \cdot 10^{-25}$ V m⁻¹ s²) [7]. We can replace in Eq. (2) the term c^2 with the electric permittivity ε_0 and the magnetic permeability μ_0 obtaining:

$$G = \frac{l_P^2}{4 \pi e \xi \varepsilon_0 \mu_0} \quad (3).$$

Eq. (3) confirms that the 4-D SQS electromagnetic properties are defining the gravitational constant.

2. SQS as the unified field model

Time-invariant superfluid quantum space (SQS) has a general n-dimensional complex structure $\mathbb{C}^n$; every point of it has complex coordinates:

$$z_i = x_i + i\, y_i \quad (4).$$

(x_i, y_i) $(i = 1, \ldots, n)$ is an ordered n-uple of real numbers $((x_i, y_i) \in \mathbb{R}^n)$; for the purpose of this article, we consider its subset $\mathbb{C}^4$ where all elementary particles are different structures of $\mathbb{C}^4$-SQS and have four complex dimensions z_i. In $\mathbb{C}^n$-SQS the elapsed time of a given material change, i.e., motion is the sum of Planck times t_P:

$$t = t_{P1} + t_{P2} + \ldots + t_{PN} = \sum_{i=1}^{N} t_{Pi} \quad (5)\,[8].$$

$\mathbb{C}^n$-SQS is time-invariant in the sense that time is not its fourth dimension. Material changes run in time-invariant $\mathbb{C}^n$-SQS and time is their duration. We do not have any experimental data that time is the fourth dimension of space and we suggest in this article a novel model where time is only the duration of change in time-invariant complex space. $\mathbb{C}^n$-SQS is the physical origin of the universal space. We call it "four-dimensional complex superfluid quantum space" ($\mathbb{C}^4$-SQS). Subatomic particles are different structures of $\mathbb{C}^4$-SQS; atoms, made out of subatomic particles, are three-dimensional physical objects, described by real geometry $\mathbb{R}^3$ and therefore follow the 3-D Euclidean geometry. Because of that we cannot fully grasp the complex subatomic level with 3-D apparatuses (Figure 1).

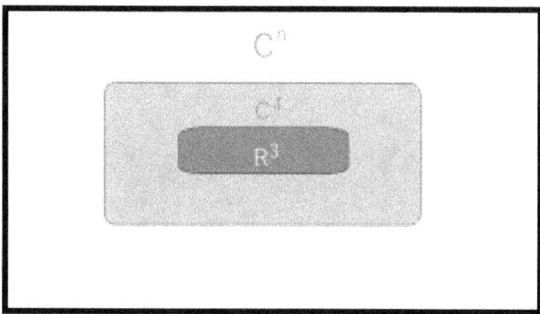

Figure 1: Structure of the $\mathbb{C}^n$-SQS universe.

The 4-D complex superfluid quantum space $\mathbb{C}^4$-SQS is the theoretical frame for the unification of gravity and the other three fundamental forces which have already been unified by the Standard Model. In this complex superfluid quantum space, we have four spatial coordinates which have a real and imaginary component. The energy density of $\mathbb{C}^4$-SQS is calculated in the terms of $\mathbb{R}^3$ matter in units kg/m³ and related to the mass m of a given physical object; every physical object with mass m is decreasing the energy density ρ_{Emin} of $\mathbb{C}^4$-SQS in its centre exactly for the amount of its energy:

$$E = mc^2 = (\rho_{EP} - \rho_{Emin})V \quad (6)$$

where ρ_{EP} is the energy density of SQS faraway of a stellar object in the interstellar space and V is its volume [6]. By Eq. (6) we can calculate the minimal energy density of space in the centre of a given physical object:

$$\rho_{Emin} = \rho_{EP} - \frac{mc^2}{V} \quad (7).$$

Eq. (7) holds from the proton scale to black holes scale. Going away from the centre of a given physical object, the energy density of space is increasing by the following equation [9]:

$$\rho_{Emin} = \rho_{EP} - \frac{3mc^2}{4\pi(r+d)^3} \quad (8)$$

where r is the radius of the physical object and d is the distance from its centre. When d tends to the infinite, ρ_{Emin} tends to ρ_{EP} (Figure 2).

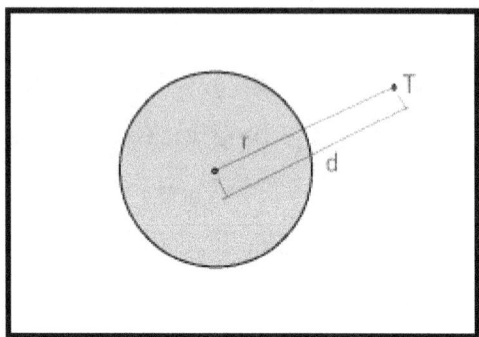

Figure 2. Energy density of $\mathbb{C}^4$-SQS in the point T at the distance d from its centre.

Gravity is carried by external pressure of $\mathbb{C}^4$-SQS towards the centre of physical objects; it is the force of $\mathbb{C}^4$-SQS pressure from the maximum energy density of $\mathbb{C}^4$-SQS towards its decreased energy density in the centre of the given physical object. Two physical objects are creating decreased area of $\mathbb{C}^4$-SQS energy density, causing outer pressure of $\mathbb{C}^4$-SQS towards its lower inner pressure. This outer pressure is gravity (Figure 3) [6].

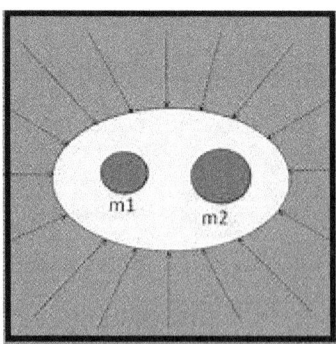

Figure 3. Gravity is the space pressure towards the physical objects.

That gravity is generated by the pressure force of the medium in which physical objects exist is also proposed by Arminjon back in 1997: "The theory starts from a tentative interpretation of gravity as Archimedes' thrust exerted on matter at the scale of elementary particles by an imagined perfect fluid ("ether"): the gravity acceleration is expressed by a formula in which the "ether pressure" ρe plays the role of the Newtonian potential. The instantaneous propagation of Newtonian gravity is obtained with an incompressible ether, giving a field equation for ρe" [9]. In our model, the medium is not filling universal space in the sense as ether should fill the universal space. Universal space itself is the physical medium in which physical objects exist. NASA has measured back in 2014 that universal space has a Euclidean shape: "Recent measurements (c. 2001) by a number of ground-based and balloon-based experiments, including MAT/TOCO, Boomerang, Maxima, and DASI, have shown that the brightest spots are about 1 degree across. Thus, the universe was known to be flat to within about 15% accuracy prior to the WMAP results. WMAP has confirmed this result with very high accuracy and precision. We now know (as of 2013) that the universe is flat with only a 0.4% margin of error" [10]. This puts in question the curvature of the universal space as suggested in General Relativity. We suggested in our previous research that the curvature of space in General Relativity (GR) is the mathematical description of some fundamental physical property of the universal space that is its variable energy density. With the increasing of space curvature in GR, the energy density of space is decreasing in our model [11].

We will use this Eq. (7) to calculate the energy density of space in the centre of different stellar objects, considering that these objects are non-rotating. In Table 1 there is the comparation of the energy densities of space in the centre of the black hole with the mass of the Sun, in the centre of the proton, in the centre of the Moon, Earth, and Sun:

Table 1. Comparation values of the minimal energy density of space with respect to the centre of indicated objects.

Centre of objects	$\rho_p = 4.64 \cdot 10^{113} Jm^{-3}$
Black hole with mass of the Sun	$\rho_p - 1.58 \cdot 10^{36} Jm^{-3}$
Proton	$\rho_p - 5.43 \cdot 10^{34} Jm^{-3}$
Earth	$\rho_p - 4.94 \cdot 10^{20} Jm^{-3}$
Moon	$\rho_p - 3.00 \cdot 10^{20} Jm^{-3}$
Sun	$\rho_p - 1.26 \cdot 10^{20} Jm^{-3}$

Hawking has proposed that proton could be a min black hole [12]. Voyager did not find these primordial black holes [13]. Our research confirms, in the centre of a proton, the minimal energy density of SQS is not low enough for a proton to become a black hole. The energy density of space in the proton centre is for the order of 10^2 higher than in the black hole centre with mass of the Sun. In the centre of a proton, the minimal energy density of SQS is for the order of 10^{14} lower than in the centre of Sun, Earth and Moon, because these stellar objects are made out of atoms where there is a vast empty space between the nucleus and electrons orbits. Proton's mass is very small compared with the mass of the Sun, but it diminishes the energy density of an extremely small area of space compared with that of Sun, that diminishes the energy density of an extremely big area of universal space; that's why the gravity force of the Sun has such a long-range.

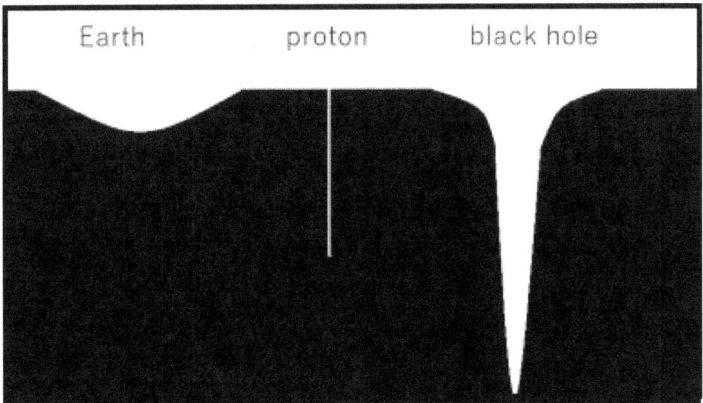

Figure 4: Energy density of $\mathbb{C}^4$-SQS in the centre of Earth, proton and black hole
(figure is an approximation)

Proton has much lower energy density of $\mathbb{C}^4$-SQS in its centre than the Earth (Figure 4); however, it has almost no attraction force because of its extremely small mass. The calculation of attraction force because of lower energy density of $\mathbb{C}^4$-SQS in the centre of proton and neutron in deuterium nucleus is as follows:

$$F_g = \frac{1.67 \cdot 10^{-27} kg \cdot 1.67 \cdot 10^{-27} kg \cdot 6.67 \cdot 10^{-11} m^3 kg^{-1} s^{-2}}{(2 \cdot 0.87 \cdot 10^{-15} m)^2}$$
$$= 6.144 \cdot 10^{-35} N$$

where $1.67 \cdot 10^{-27}$ kg is the mass of proton and neutron, and $0.87 \cdot 10^{-15}$ m is their radius. This calculation confirms that gravity force and strong nuclear force are not the same force as suggested by Vayenas and Souentie [14]. The binding energy between a proton and neutron in deuterium nucleus is $2224575 \pm 9 eV$ which is $3.564 \cdot 10^{-13} J$ [15]. The error of their proposal is of rate 10^{22}.

From the macro to the microscale, it holds that a given physical object is interacting with the $\mathbb{C}^4$-SQS in which it exists and the result of this interaction are the inertial mass m_i and the gravitational mass m_g:

$$m_i = m_g = \frac{(\rho_{EP} - \rho_{Emin})V}{c^2} \quad (9).$$

The rest mass m_0 of the proton is not its inertial mass m_i, but is related to the amount of $\mathbb{C}^4$-SQS energy E which is incorporated in the proton, as per Eq. (10):

$$m_0 = \frac{E}{c^2} \quad (10).$$

The inertial mass m_i of the proton is the result of proton interaction with the $\mathbb{C}^4$-SQS energy and the decrease in energy density of $\mathbb{C}^4$-SQS in proton's centre is exactly for the amount of its mass and volume V as we can see in Eq. (9). This decreased energy density of $\mathbb{C}^4$-SQS is the physical origin of proton's inertial mass. Einstein has proved that inertial mass and gravitational mass of a given physical object are equal. We have shown that they have the same origin in the decreased energy density of $\mathbb{C}^4$-SQS accordingly to the Eq. (9).

That inertial mass and gravitational mass are one phenomenon ("to be identical thing") was also proposed by Rueda and Haisch back in 2005: "There are additional consequences that make this approach of assuming real interactions between the electromagnetic quantum vacuum and matter appear promising. It can be shown that the weak principle of equivalence — the equality of inertial and gravitational mass 1 — naturally follows. In the quantum vacuum inertia hypothesis, inertial and gravitational mass are not merely equal, they prove to be the identical thing" [16].

The strong nuclear force is carried by gluons which bind together quarks inside the proton and neutron. Residual nuclear force between quarks is acting also outside protons and neutrons and hold them together in nucleus of an atom. In the model here presented gluons are excitations of $\mathbb{C}^4$-SQS and they represent 99% of proton mass. In this perspective, the proton mass can be seen as the excitation of the $\mathbb{C}^4$-SQS in the form of gluons and quarks. In Meis model the mass of electron m_e and mass of proton M_p are expressed by the physical properties of the electromagnetic vacuum:

$$m_e = 2\pi c e^2 \left|\frac{\xi}{\mu_B}\right| = 9.109 \cdot 10^{-31} \, kg \quad (11)$$

$$M_P = 2\pi c e^2 \left|\frac{\xi}{\mu_B}\right| = 1.672 \cdot 10^{-27} \, kg \quad (12)$$

The mass M_I of any elementary particle I can be expressed using **Eq. (11)**:

$$m_i = 2\pi c e^2 \left|\frac{\xi}{\mu_i}\right| \quad (13)$$

with $|\mu_i| = \mu_B$ for the electron and $|\mu_i| = \left(\frac{2\alpha}{n_i}\right)\mu_B$ for other particles, with n_i an integer and α the fine structure constant, $\mu_B = 9{,}274 \cdot 10^{-27} JT^{-1}$ [7].

In Meis model, as well in $\mathbb{C}^4$-SQS model, elementary particles are different energy structures of the $\mathbb{C}^4$-SQS energy. This is also the view of Erwin Schrödinger who used to say: "What we observe as material bodies and forces are nothing but shapes and variations in the structure of space" [17]. This is expressed in Einstein formula $E = mc^2$; E is the $\mathbb{C}^4$-SQS energy which is incorporated in a given physical object, m is the mass of the object.

Relativistic particles are interacting with the $\mathbb{C}^4$-SQS and additionally integrating $\mathbb{C}^4$-SQS energy into its structure. Relativistic energy E of a given accelerated particle is the sum of the rest energy E_0 and kinetic energy E_K which is incorporated energy of $\mathbb{C}^4$-SQS due to the motion of the particle [18]:

$$E = E_0 + E_K = \gamma m_0 c^2 = (\rho_{EP} - \rho_{EminR})V \quad (14),$$

where γ is Lorentz factor, m_0 is proton rest energy, ρ_P is Planck energy density, ρ_{EminR} is additionally diminished energy density of $\mathbb{C}^4$-SQS in the centre of the proton because the proton is additionally absorbing $\mathbb{C}^4$-SQS energy and so increasing its mass and energy, V is the volume of the proton at rest.

The unification of electromagnetism and weak nuclear force into electroweak force was independently proposed by Sheldon Glashow, Abdus Salam and Steven Weinberg in the sixties of the last century; we introduced here a model where all four fundamental forces are carried by $\mathbb{C}^4$-SQS. Gravity force is carried by variable density of $\mathbb{C}^4$-SQS, strong nuclear force and electroweak force are carried by the excitation of $\mathbb{C}^4$-SQS.

In the model of $\mathbb{C}^4$-SQS the electric field is the excitation of $\mathbb{C}^4$-SQS on the three real dimensions X_1, X_2, X_3, and the magnetic field is the excitation of $\mathbb{C}^4$-SQS on the tree real dimension X_2, X_3, X_4. Both fields have in common dimensions X_2 and X_3. The photon is then the excitation of $\mathbb{C}^4$-SQS on X_1, X_2, X_3, X_4 dimensions, it is a 4-D wave of $\mathbb{C}^4$-SQS; the light has a constant speed for all moving observers because it is a wave of $\mathbb{C}^4$-SQS. When the source of light is moving closer to the observer or moving away from the observer the frequency of light will respect the Doppler effect. The source of light and the moving observer are all moving in the $\mathbb{C}^4$-SQS. This model explains the physical meaning of the first postulate of Special Relativity, i.e., that the light has the same velocity for all observers; the light is a 4-D wave of $\mathbb{C}^4$-SQS in which the observer and the 3-D source of light are moving. The motion of the observer or the motion of the light source creates the Doppler effect but the light speed remains unchanged. In Special Relativity the photon is moving in a 4-D space of Minkowski, where time t is the element of the fourth dimension $X_4 = ict$. We have shown that time is just the numerical sequential order of material changes, i.e., motion in $\mathbb{C}^4$-SQS. When we measure the numerical order of photon motion from the point A to the point B in $\mathbb{C}^4$-SQS, we get duration. The photon is the wave of $\mathbb{C}^4$-SQS and does not move in some physical space-time, it moves in time-invariant universal space [19].

4. Complex time-invariant $\mathbb{C}^4$-SQS is the medium of quantum entanglement

Time-invariant superfluid quantum space is the medium of entanglement EPR-type [19,20]. In this perspective, time as duration of material change, i.e., motion in $\mathbb{C}^4$-SQS, can be seen as an emergent property of entangled universe. Time as duration enters existence when measured by the observer [8,19]. Moreva et al. came to the same conclusion, namely time is an emergent property of entanglement, starting their research from a different perspective [21].

Einstein has interpreted the time t as the 4th coordinate X_4 of a Minkowski manifold. He wrote: "If we replace $x, y, z, \sqrt{-1}\,ct$ by x_1, x_2, x_3, x_4, we also obtain the result that $ds^2 = dx_1^2 + dx_2^2 + dx_3^2 + dx_4^2$ is independent of the choice of the body of reference. We call the magnitude ds the "distance" apart of two events or four-dimensional points. Thus, if we choose as time variable the imaginary variable $\sqrt{-1}\,ct$ instead of the real quantity t, we can regard the continuum space-time, in accordance with the special theory of relativity, as an "Euclidean" four-dimensional continuum, a result following by the consideration of the preceding section" [22]. In the above citation, Einstein suggestion that we can choose the time variable t as the imaginary variable can be written as follows:

$$t = \sqrt{-1}\,ct \quad (15).$$

Eq. (15) is false because on the left side of the equation we have t and on the right side we have $\sqrt{-1}\,ct$. Combining Eq. (6) with equation well know equation $X_4 = ict$ we get:

$$X_4 = itc^2\sqrt{-1} \qquad (16).$$

Eq. (16) confirms that Einstein did a mistake keeping and interpreting time as the dimension of a four-dimensional continuum. Physics is still today suffering this misinterpretation of time that is solved in this article: time is the duration of a given physical object motion in time-invariant space. There is no "distance in time" in the universe because there is no physical time in which events happen. The universal changes run in time-invariant space which means that the entire universe is existing simultaneously. This is so-called "absolute simultaneity" in which there is no physical past and there is no physical future [19,20].

Several authors are proposing that entanglement is induced by gravity [23,24,25]. On the other hand, there is a proposal that entanglement influences gravity: "To summarize, we have shown that entanglement can affect the gravitational field. This suggests that entanglement "has a weight". The perturbations in the gravitational field depend on the amount of entanglement and vanish for vanishing quantum correlations" [26]. We have shown in this article that gravity and entanglement are carried by the same medium which is SQS. In the model presented in this article gravity force between two physical objects does not induce entanglement and entanglement has no impact on gravity. Two entangled physical objects are entangled via SQS which variable energy density is also carrying gravity. Gravity does not influence entanglement and vice versa is also valid. In the model here presented gravity and entanglement are both induced by the superfluid quantum space that is time-invariant [19,20].

5. Conclusions

The unified field theory of Albert Einstein is one of the main goals of modern physics. This goal can be achieved by the development of complex $\mathbb{C}^4$-SQS as the fundamental arena of the universe. Elementary particles and consequently strong nuclear force and electroweak force forces are different structures of $\mathbb{C}^4$-SQS. Gravity does not require the existence of some hypothetical particle graviton. It is carried directly by the variable energy density of time-invariant complex $\mathbb{C}^4$-SQS that is the medium of quantum entanglement EPR-type experiments.

References:

1. Sbitnev, V.I. Hydrodynamics of the Physical Vacuum: II. Vorticity Dynamics. Found Phys 46, 1238–1252 (2016). **https://doi.org/10.1007/s10701-015-9985-3**.

2. Sbitnev, V.I. Hydrodynamics of the Physical Vacuum: I. Scalar Quantum Sector. Found Phys 46, 606–619 (2016). **https://doi.org/10.1007/s10701-015-9980-8**.

3. Bezuglov, M, False vacuum decay in quantum mechanics and four dimensional scalar field theory, EPJ Web of Conferences, 177, 09001 (2018). **https://doi.org/10.1051/epjconf/201817709001**.

4. Planck Collaboration, Planck 2015 results. XIII. Cosmological parameters, Astronomy & Astrophysics, 594, A13 (2016). arXiv:1502.01589, doi:10.1051/0004-6361/201525830.

5. Peebles, P.J.E., Open problems in cosmology, Nuclear Physics B - Proceedings Supplements, 138, 5-9 (2005).

6. Fiscaletti, D., & Šorli, A.S., Quantum Relativity: Variable Energy Density of Quantum Vacuum as the Origin of Mass, Gravity and the Quantum Behaviour, Ukrainian Journal of Physics, 63(7), 623 (2018). **https://doi.org/10.15407/ujpe63.7.623**.

7. Meis C., Primary Role of the Quantum Electromagnetic vacuum in Gravitation and Cosmology (2020). doi: 10.5772/intechopen.91157.

8. 8. Fiscaletti, D., Šorli, A. Perspectives of the Numerical Order of Material Changes in Timeless Approaches in Physics. Foundations of Physics, 45, 105-133 (2015). **https://doi.org/10.1007/s10701-014-9840-y**.

9. Arminjon M. Scalar theory of gravity as a pressure force, Revue Roumaine des Sciences Techniques - Mécanique Appliquée, 42, No. 1-2, pp. 27-57 (1997) **https://arxiv.org/ftp/arxiv/papers/0709/0709.0408.pdf**

10. NASA, (2014). **https://wmap.gsfc.nasa.gov/universe/uni_shape.html**

11. Fiscaletti D., Sorli A. Space-Time Curvature Of General Relativity And Energy Density Of A Three-Dimensional Quantum Vacuum, Annales Universitatis Mariae CurieSklodowska: Physica, Sectio AAA; Lublin Vol. 69, Iss. 1, (2015): 53- 78. DOI:10.1515/physica-2015-0004

12. Hawking, S., Gravitationally Collapsed Objects of Very Low Mass, Monthly Notices of the Royal Astronomical Society, 152(1), 75-78 (1971).
https://doi.org/10.1093/mnras/152.1.75.

13. Boudaud, M. & Cirelli, M. Voyager 1 e ± Further Constrain Primordial Black Holes as Dark Matter, Physical Review Letters, 122, 041104 (2019).
https://journals.aps.org/prl/abstract/10.1103/PhysRevLett.122.041104.

14. Vayenas C.G., Souentie S.NA. (2012). Force Unification: Is the Strong Force Simply Gravity?. In: Gravity, Special Relativity, and the Strong Force. Springer, Boston, MA.
https://doi.org/10.1007/978-1-4614-3936-3_12.

15. C.Van Der Leun, C.Alderliesten, The deuteron binding energy, Nuclear Physics A, Volume 380, Issue 2 (1982). Pages 261-269 https://doi.org/10.1016/0375- 9474(82)90105-1 11

16. Rueda A., Haisch B., Gravity and the quantum vacuum inertia hypothesis, Volume 14, Issue 8 (2005).
https://doi.org/10.1002/andp.200510147

17. Schrödinger, E., Space-Time Structure, Cambridge: Cambridge University Press (1985).

18. Relativity Reborn: Based on Bijective Physis, Amazon (2019), ISBN-10: 1687725888

19. Šorli, A., & Čelan, Štefan. The End of Space-time: Physics-Mathematics. International Journal of Fundamental Physical Science, 10(4), 31-34. (2020).
https://doi.org/10.14331/ijfps.2020.330139

20. Fiscaletti, D., Šorli, A.S., The Infinite History of NOW: The Timeless Background of Contemporary Physics, Nova Science Publishers (2014). ISBN: 978-1-63117-283-0

21. Moreva, E., Brida, G., Gramegna, M., Giovannetti, V., Maccone, L., and Genovese, M., Time from quantum entanglement: An experimental illustration, Physical Review A, 89, 052122 (2014).
https://doi.org/10.1103/PhysRevA.89.052122.

22. Einstein, A., Relativity: The Special and General Theory, Methuen & Co Ltd, p.93 (1916).

23. Krisnanda, T., Tham, G.Y., Paternostro, M. et al. Observable quantum entanglement due to gravity. npj Quantum Inf 6, 12 (2020). **https://doi.org/10.1038/s41534-020-0243-y**.

24. Bose, S. et al. Spin entanglement witness for quantum gravity. Phys. Rev. Lett. 119, 240401 (2017). **https://doi.org/10.1103/PhysRevLett.119.240401**.

25. Marletto, C. & Vedral, V. Gravitationally induced entanglement between two massive particles is sufficient evidence of quantum effects in gravity. Phys. Rev. Lett. 119, 240402 (2017).
https://doi.org/10.1103/PhysRevLett.119.240402.

26. David Edward Bruschi, On the weight of entanglement, Physics Letters B, Volume 754, 10 March 2016, Pages 182-186, **https://doi.org/10.1016/j.physletb.2016.01.034**

Schwarzschild energy density of superfluid quantum space and mechanism of AGNs' jets

Abstract

Active galactic nuclei (AGNs) are throwing in the interstellar space huge jets of energy in the form of elementary particles. The calculation of the energy density of space in the centre of the black hole with the mass of the Sun shows that in the space-time singularity of such a black hole energy density of space there is so low that atoms become unstable and fall apart in elementary particles. In this sense, AGN is a rejuvenating system of the universe. It transforms its own old matter into fresh energy in the form of jets.

Keywords: Space-time singularities; Energy density of quantum vacuum; AGN; Jets.

1. Introduction

Several pieces of research suggest that superfluid quantum vacuum also named superfluid quantum space (SQS) is the physical origin of the universal space [1,2,3,4]. The idea of space-time as the fundamental arena of the universe is replaced by the idea that universal space is a type of energy that has superfluid properties. One of these superfluid properties is that every physical object is diminishing the Planck energy density ρ_{EP} of the superfluid quantum space which is the origin of the universal space in its centre exactly for the amount of its mass m and energy E:

$$E = mc^2 = (\rho_{EP} - \rho_{Ec})V \quad (1),$$

where ρ_{min} is the energy density of the universal space in the centre of the physical object and V is the volume of the object [3]. The "no hair theorem" states that a black hole can be defined by three parameters: mass, electric charge, angular momentum [5]. Considering the variable energy density of universal space, I introduce a new parameter, the "minimal energy density of SQS in the centre of a black hole". By Eq. (1), we get:

$$\rho_{Ec} = \rho_{EP} - \frac{mc^2}{V} \quad (2)$$

where ρ_{Ec} is the energy density of SQS in the centre of a black hole, m is the mass of the black hole and V is its volume.

2. Calculation of the "Schwarzschild energy density"

"Schwarzschild energy density" one can calculate using Eq. (2):

$$\rho_{E.Sch.} = \rho_{EP} - \frac{3m_\odot c^2}{4\pi r_{Sch.}^3},$$

where $m_\odot$ is the mass of the Sun, and its correspondent Schwarzschild radius $r_{Sch.}$ is $3 \cdot 10^3 m$.

$$\rho_{E.Sch.} = 4.64 \cdot 10^{113} Jm^{-3} - 1.58 \cdot 10^{36} Jm^{-3}.$$

When in the centre of the stellar object the value of energy density of SQS ρ_{Ec} is smaller as Schwarzschild energy density, the atoms in the centre become unstable and are falling apart into elementary particles:

$$\rho_{Ec} < \rho_{E.Sch.} \rightarrow atoms\ are\ unstable$$

The Schwarzschild energy density offers a new interpretation of space-time singularities in the centre of a black hole: "If, as seems justifiable, actual physical singularities in space-time are not to be permitted to occur, the conclusion would appear inescapable that inside such a collapsing object at least one of the following holds: (a) Negative local energy occurs. (b) Einstein's equations are violated. (c) The space-time manifold is incomplete. (d) The concept of space-time loses its meaning at very high curvature – possible because of quantum phenomena. In fact (a), (b), (c), (d) are somewhat interrelated, the distinction being partly one of attitude of mind" [6]. We suggest that space-time singularity in the centre of black hole indicate that in the centre of a black hole exist critical physical circumstances that we previously defined as "the energy density of SQS e is below the Schwarzschild energy density $\rho_{E.Sch.}$".

According to Newton's Shell theorem in space-time singularity, gravity force does not tend to the infinite value; it tends to zero. Going inside the black hole at the distance d from the surface towards the centre gravity force on a given object with mass m is diminishing regarding the gravity force on the surface according to the Eq. (3):

$$F_{g1} = \frac{mM_1 G}{r_1} \quad (3),$$

where m is the mass of a given object, M_1 is the mass of the black hole shell with the radius r_1, and G is gravitational constant. When r_1 is tending to the zero, M_1 is also tending to the zero, and gravity force F_{g1} is also tending to the zero:

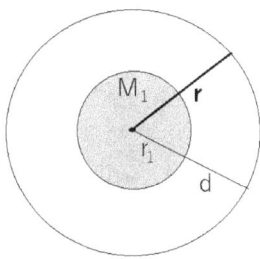

Figure1: Gravity force inside black hole

In the centre of the black hole, there is no gravity force. The extreme physical circumstance in the centre of the black hole is that the energy density of SQS there is below Schwarzschild energy density. This model is adding to the understanding of the interior of black holes which back in 2000 was still an open question: "We have confined ourselves in this paper to a specific example that relaxes the condition of asymptotic flatness while preserving time-symmetry. The starting point here is the static black hole in the Einstein universe which belongs to the family of solutions presented by Vaidya. In this spacetime the black hole is well defined as the Killing horizon. However, the nature of the interior of the black hole is not entirely clear" [7].

3. Variable energy density of SQS at the distance d from the centre of the stellar objects

The energy density of the universal space ρ_{Ed} at the distance d from the centre of a given stellar object with mass m and radius r is calculated using the equation below that is developed on the basis of the Eq. (2):

$$\rho_{Ed} = \rho_{EP} - \frac{3mc^2}{4\pi(r+d)^3} \quad (4).$$

When d is zero, Eq. (4) is equal to the Eq. (2), when d tents towards the infinite, energy density ρ_{Ed} tends towards Planck energy density ρ_{EP}. We will use this formula to calculate the energy density of SQS in the centre of different stellar objects, considering that these objects are non-rotating. In Table 1 there is the comparation of the energy densities of SQS in the centre of the black hole with the mass of the Sun, in the centre of the proton, in the centre of the Moon, Earth, and Sun:

Table 1. Comparation values of the minimal energy density of space with respect to the centre of indicated objects.

Centre of objects	$\rho_{EP} = 4.64 \cdot 10^{113} Jm^{-3}$
Black hole with mass of the Sun	$\rho_{EP} - 1.58 \cdot 10^{36} Jm^{-3}$
Proton	$\rho_{EP} - 5.45 \cdot 10^{34} Jm^{-3}$
Earth	$\rho_{EP} - 4.97 \cdot 10^{20} Jm^{-3}$
Moon	$\rho_{EP} - 3.01 \cdot 10^{20} Jm^{-3}$
Sun	$\rho_{EP} - 1.27 \cdot 10^{20} Jm^{-3}$

In the centre of a proton, the minimal energy density of SQS is for the order of 10^2 too high for the proton to become a mini black hole as proposed by Hawking [8]. Voyager did not discover mini black holes in interstellar space [9]. The energy density of SQS in the proton centre is lower than in the centre of Sun, Earth and Moon because these stellar objects are made out of atoms where there is a vast empty space between the nucleus and electrons orbits. Proton's mass is very small compared with the mass of the Sun, but it diminishes the energy density of an extremely small area of space compared with that of Sun, that diminishes the energy density of an extremely big area of universal space; that's why the gravity force of the Sun has such a long-range.

We will compare the energy density of SQS of a stationary black hole with the mass of the Sun and energy density of SQS of the Sun at given distances from the centre (Table 2).

Table 2. Comparation values of the minimal energy density of SQS with respect to the distance by centre of indicated objects.

Sun centre	**Black hole** centre
$\rho_{EP} - 1.27 \cdot 10^{20} Jm^{-3}$	$\rho_{EP} - 1.58 \cdot 10^{36} Jm^{-3}$
Distance from the centre	Distance from the centre
10^2 km	10^2 km
$\rho_{EP} - 1.27 \cdot 10^{20} Jm^{-3}$	$\rho_{EP} - 3.91 \cdot 10^{31} Jm^{-3}$
10^3 km	10^3 km
$\rho_{EP} - 1.26 \cdot 10^{20} Jm^{-3}$	$\rho_{EP} - 4.24 \cdot 10^{28} Jm^{-3}$
10^4 km	10^4 km
$\rho_{EP} - 1.22 \cdot 10^{20} Jm^{-3}$	$\rho_{EP} - 4.27 \cdot 10^{25} Jm^{-3}$
10^5 km	10^5 km
$\rho_{EP} - 8.48 \cdot 10^{19} Jm^{-3}$	$\rho_{EP} - 4.28 \cdot 10^{22} Jm^{-3}$
10^6 km	10^6 km
$\rho_{EP} - 8.77 \cdot 10^{18} Jm^{-3}$	$\rho_{EP} - 4.28 \cdot 10^{19} Jm^{-3}$
0.1 AU	0.1 AU
$\rho_{EP} - 1.11 \cdot 10^{16} Jm^{-3}$	$\rho_{EP} - 1.28 \cdot 10^{17} Jm^{-3}$
0.5 AU	0.5 AU
$\rho_{EP} - 9.94 \cdot 10^{13} Jm^{-3}$	$\rho_{EP} - 1.02 \cdot 10^{14} Jm^{-3}$
1 AU	1 AU
$\rho_{EP} - 1{,}26 \cdot 10^{13} Jm^{-3}$	$\rho_{EP} - 1.28 \cdot 10^{13} Jm^{-3}$
10 AU	10 AU
$\rho_{EP} - 1{,}27 \cdot 10^{10} Jm^{-3}$	$\rho_{EP} - 1.28 \cdot 10^{10} Jm^{-3}$

Going from the centre of the black hole, the energy density of the SQS is increasing at a much higher rate than going away from the centre of the Sun. At the distance of 1 AU from the centre of both stellar objects, the energy density of SQS is at the same rate comparing the Planck energy density and is increasing by the same values with the increase of the distance.

4. The quantum mechanism of AGNs' jets

In the centre of black holes, atoms are transforming back into elementary particles. This creates enormous pressure and if gravity pressure of the black hole is not big enough, such a black hole explodes in a supernova [10]. When the black hole gravity pressure is strong enough, as it is the case for example with the black hole in the quasar SMSSJ215728.21–360215.1 which has about $(3.4 \pm 0.6) \cdot 10^{10}\ M_\odot$ [11], the transformation of matter into elementary particles creates the explosion that opens the hole in the direction of the rotational axis (Figure 2).

Figure 2: Cross-section of a black hole in the centre of the quasar SMSSJ215728.21–360215.1

Through this hole, in the direction of rotation, the black hole is throwing a jet of elementary particles into the intergalactic SQS (Figure 2).

Figure 3: Jets of a black hole in the centre of a galaxy (with permission of Southern European observatory).

Centres of AGN's where energy density of SQS is lower than Sch. energy density are mechanisms where the matter falls apart into elementary particles and forms jests. We give in this article a solution to the mystery of jets production following Einstein's idea that matter can be transformed into energy and vice versa: "Relativistic magnetized jets from active galaxies are among the most powerful cosmic accelerators, but their particle acceleration mechanisms remain a mystery" [12]. These jets are building material for new stars formation; black holes are then rejuvenating systems of the universe: "old" matter is transformed into "fresh" energy in the form of AGNs jets.

7. Conclusions

The law of energy conservation requires that AGN's jets must have some physical sources. It is shown in this article that these jets are originated in the process of matter falling apart in the centres of

AGNs, where there are space-time singularities and energy density of SQS is below the Schwarzschild energy density.

References:

1. Sbitnev, V.I. Hydrodynamics of the Physical Vacuum: II. Vorticity Dynamics. *Found Phys* **46,** 1238–1252 (2016). **https://doi.org/10.1007/s10701-015-9985-3**.

2. Sbitnev, V.I. Hydrodynamics of the Physical Vacuum: I. Scalar Quantum Sector. *Found Phys* **46,** 606–619 (2016). **https://doi.org/10.1007/s10701-015-9980-8**.

3. 3. Fiscaletti, D., & Sorli, A.S. Quantum Relativity: Variable Energy Density of Quantum Vacuum as the Origin of Mass, Gravity and the Quantum Behaviour. **Ukr. J. Phys. 63**(7), 623 (2018). **https://doi.org/10.15407/ujpe63.7.623.**

4. Šorli, A.S. Mass–Energy Equivalence Extension onto a Superfluid Quantum Vacuum. *Sci Rep* **9,** 11737 (2019). **https://doi.org/10.1038/s41598-019-48018-2**.

5. **Misner, C.W., Thorne, K.S. Wheeler, J.A**. *Gravitation*. San Francisco: **W.H. Freeman,** *875-876 (1973).*

6. Penrose, R. Gravitational Collapse and Space-Time Singularities. *Phys. Rev. Lett.* **14**, 57 (1965). **https://doi.org/10.1103/PhysRevLett.14.57**.

7. Nayak, K.R., Mac Callum, M.A.H., and Vishveshwara, C.V. Black Holes in Non-flat Backgrounds: the Schwarzschild Black Hole in the Einstein Universe (2000). **https://arxiv.org/pdf/gr-qc/0006040v1.pdf**.

8. Hawking, S. W. Gravitationally collapsed objects of very low mass. MNRAS **152**, 7 (1971). **https://doi.org/10.1093/mnras/152.1.75**.

9. Boudaud, M. & Cirelli, M. Voyager 1 e± Further Constrain Primordial Black Holes as Dark Matter. *Phys. Rev. Lett.* **122**, 041104 (2019). **https://doi.org/10.1103/PhysRevLett.122.041104**.

10. Croswell K., Inner Workings: A massive star dies without a bang, revealing the sensitive nature of supernovae, PNAS January 21, 2020 117 (3) 1240-1242; **https://doi.org/10.1073/pnas.1920319116**.

11. Onken, C.A., Bian, F., Fan, X., Wang, F., Wolf, C., Yang, J. A thirty-four billion solar mass black hole in SMSS J2157-3602, the most luminous known quasar. MNRAS **496**(2), 2309-2314 (2020). **https://doi.org/10.1093/mnras/staa1635**.

12. Alves E.P., Zrake, J., Fiuza, F. Efficient Nonthermal Particle Acceleration by the Kink Instability in Relativistic Jets, Phys. *Rev. Lett.* **121**, 245101 (2018). **https://arxiv.org/abs/1810.05154**.

Multiverse in permanent Equilibrium

Abstract

In the time-invariant universe model, material changes run in time-invariant superfluid quantum space (SQS). Changes have no duration on their own. Time as the duration of changes enters existence only when measured by the observer. Time-invariant superfluid quantum space (SQS) has a variable energy density that defines the velocity of material changes. More SQS is dense, faster is the velocity of changes. Every physical object is diminishing energy density of SQS exactly for the amount of its energy E and correspondent mass m. In the centre of AGNs' energy density of SQS is so low that atoms become unstable. They fall apart into elementary particles in the form of jets. These jets are the raw material for the formation of new stars. AGNs' are rejuvenating systems of the universe that is non-created and eternal.

Keywords: universe, superfluid quantum space, CMB, cosmological redshift, AGN.

1. Introduction

The result of several pieces of research is that the superfluid quantum vacuum also named superfluid quantum space (SQS) is the physical origin of the universal space, the fundamental arena of the universe [1,2]. Considering that with clocks we measure the duration of a material change, i.e., motion in SQS one can conclude that time is not the fourth dimension of SQS. We do not have a shred of single experimental evidence that time is the 4th dimension of the universal space and it is time that in physics we pustulate time on the basis of its measurement; time is the duration of the change in time-invariant universal space. This means that time as the duration in order to exist needs to be measured from the side of the observer There is no physical past nor physical future in the universe, a time when measured, is a duration of the change in time-invariant universal space [3,4]. Considering that universal space is time-invariant, it means that we cannot study the universe from the perspective of a system that runs in some physical time. The universe is time-invariant in the sense that time is not the dimension in which the universe would run. Changes in the universe are irreversible, when change X+1 enters existence, the change X is not existent anymore. When change X+2 enters the existence the change X+1 is not existent anymore. The variable energy density of SQS defines the velocity of material changes, rate of clocks included. The rate of clocks is at the maximum in interstellar space where the energy density of SQS is at the maximum and is diminishing with the decreasing of the energy density of SQS on the surface of a given stellar object. For example, when one second has passed on the Earth surface, at the point T in infinity 1.000000000695915 second has passed [5].

    The Big Bang model is based on some hypothetical physical past events as the initial explosion and inflation that have not been directly observed. The time-invariant space is confirming the principle of time translation symmetry (TTS) which states that laws of physic are independent of time [4]. In the time-invariant universe there is no physical past and there is no physical future. Changes run in time-invariant SQS. The idea that gravity immediately after the big bang was acting as the repulsive force has no scientific validity.

The idea of how the initial super-force with the cooling of the universe has been transformed into four fundamental forces has no direct experimental evidence.

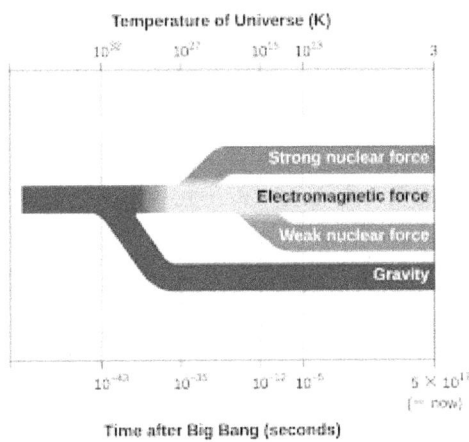

Figure 1: Hypothetical super-force transformation into four fundamental forces

The only universe that really exists is the one the observer can perceive and measure. The ideas of the beginning, inflation, and recombination are unproven speculations. We will develop in this article the cosmological model that is based only on a direct reading of astronomical data and is without ad-hoc hypothetical speculations.

2. Cosmological redshift is "tired" light effect

The redshift of the light coming from distant galaxies is today understood as the experimental proof of the universal space expansion. We do not have in scientific literature a theoretical model with mathematical evaluations that exactly predict how the light would behave when moving in the opposite direction of expanding space. This is a serious inconvenience and a puzzle that needs to be solved. The Doppler effect is observed on the Earth's surface and Earth is moving around the Sun in the stationary space. In the area of our solar system, the space is not expanding; if it would expand, this would change the trajectories of the planets orbiting around the Sun.

Recent research suggests that the SQS has the value of Planck energy density ρ_{EP} [1,6]. The gravitational constant G can be expressed with Planck energy density ρ_{EP} and Planck time t_P as: $G = \frac{1}{\frac{\rho_{EP}}{c^2} t_P^2}$. If the universe would expand, the energy density of the SQS would diminish and consequently the gravitational constant would increase. The gravitational constant was measured first back in 1798 by Henry Cavendish. Since then, the value of gravitational constant is stable, meaning that the density of SQS is also stable. This is suggesting that the universe is not expanding.

Not only the gravitational constant, also the magnetic permeability μ_0 and the electric permittivity ε_0 of the SQS are defined by its energy density. The increase and decrease of the energy density of the SQS would be a cause for the change of magnetic permeability μ_0 and electric permittivity ε_0 and would consequently change the light speed. This last was exactly measured by English astronomer James Bradley back in 1729 [7]. The constancy of μ_0, ε_0 and light speed is suggesting that the energy density of the SQS is constant and that universe is not expanding.

Stephen Hawking has predicted that the universe started by the mathematical point [8]. Back in 2014, NASA has measured with the 0,4% of error that the SQS has Euclidean shape by measurement of the sum of angles between three stellar objects and getting 180° [9]. This means that the SQS can be considered infinite in its volume. On the question how a mathematical point could extend into infinite space of the universe has no answer; we know in mathematics that the mathematical point is dimensionless and cannot be transformed into a given volume. Also, the expansion of the universe is contradictory to NASA results, meaning that the SQS does not have Riemann shape, but Euclidean shape and cannot expand. The Euclidean space, infinite in volume, cannot expand. Only finite spheric Riemann space can expand.

We have a plausible explanation of cosmological redshift. When light is coming to us from remote galaxies, it moves against the space fluctuations which are carrying gravity force. Every physical object is diminishing the Planck energy density ρ_{EP} of the superfluid quantum vacuum (SQS) that is the physical origin of the SQS in its centre exactly for the amount of its mass m and energy E:

$$E = mc^2 = (\rho_{EP} - \rho_{Emin})V \quad (1),$$

where ρ_{Emin} is the energy density of the SQS in the centre of the physical object and V is the volume of the object. Superfluid quantum space fluctuations are flowing from outer interstellar space where SQS has maximum energy density towards lower energy density of SQS in the centre of stellar objects; these SQS fluctuations are carrying gravity force [1,5,6]. Light is moving in the opposite direction of these space fluctuation and is that why losing some of its energy. The result is the cosmological redshift. Swiss astronomer Zwicky has named this effect "tired light effect" (Figure 1) [10].

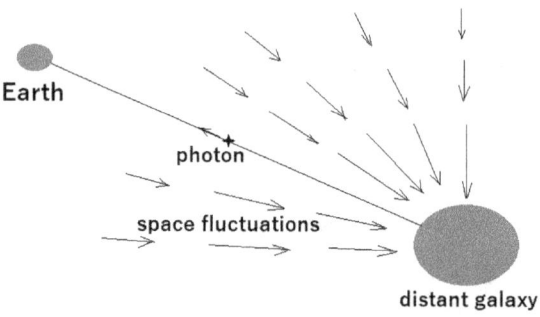

Figure 2: Light is losing some of its energy when moving in the opposite direction of the space fluctuations that carry gravity.

3. CMB is the radiation of the existent universal space

Astronomical observations confirm that the SQS is radiating uniform CMB radiation [11]. The Big Bang model suggests that the CMB radiation is the relic radiation from some remote physical past. The SQS is time-invariant; no signal can move through some hypothetical physical time; all signals move in the time-invariant space. The idea that CMB is radiation from some remote physical time is not falsifiable and should be abandoned in the name of cosmology progress. CMB has its source in present time-invariant universal space. Experimental physics is confirming a given signal we can only reach from the existent physical source, and the remote physical past is physically non-existent. Any kind of radiation must have a physical source; the remote past event cannot be this physical source. The proposal that the CMB signal is relic radiation that was created in some remote physical past and is still present is an ad-hoc proposal that was never confirmed by an experiment. The discovery of CMB absolutely does not prove the existence of a recombination period that should be existing around 380 000 after some hypothetical Big Bang. The scientific fact is that CMB is the radiation of existent universal space that has kits physical origin in SQS that is time-invariant. SQS is constantly radiating CMB and absorbing it back. In a similar way particles and antiparticles are continuously popping out of the vacuum and disappearing back into it.

4. Big Bang cosmology and Einstein's steady-state cosmology have no answer about matter creation

The radius of the mapped universe measured on the basis of astronomic observations is 46.6 billion light years. This radius is so large that, according to the age of Big Bang cosmology, the universe should expand with the velocity of 3,34 light speed to reach the mapped size of the universe [12]. We do not have any data such a velocity could exist in physical universe. The idea of a 3,34-time the light velocity has no correspondence with the physical world. The discrepancy between the measured mapped universe and the hypothetical expansion of the universe is a big unresolved question. The assumption that the universe could expand with the 3,34-time light speed is an ad-hoc theoretical prediction without the support of experimental data.

Alan Gut hypothesis was that the energy of gravity and that of matter have been multiplying in inflation period. The energy of gravity E_g is negative, the energy of matter E_m is positive, their sum is zero and in inflation on the contrary they multiply [13]. We can describe his idea mathematically as follows:

$$nE_m + (-nE_g) = 0 \quad (2).$$

Firstly, we never observed negative gravitational energy. Secondly, we never observed that energies are multiplying out of nothing. Gut's idea is against the first law of thermodynamics and is not bijective. There is no logical answer also about where both energies came into existence in the hypothetical inflation. Eq. (3) is mathematically right, but it does not fulfil the test of bijectivity, meaning that it does not correspond to some real process in physical world. The Big Bang model is not falsifiable.

The model of the universe presented in this article is based only on the obtained experimental data, is falsifiable. There are no theoretical speculations as in the case of the Big Bang model. The cosmology model presented in this article is based only on direct reading of experimental data. Thinking that the gravitational energy could be negative is logically inconsistent, because we never observed to date positive or negative energy in the universe. We know that there are precise conventions on the sign of energy, conventions adopted in all areas of physics, such as thermodynamics (absorbed energy = positive; energy released energy = negative). But these are adopted conventions, no one has ever measured that energy has an associated mathematical sign. This is also in line with the principle of bijectivity introduced in the article. Also, the idea that the energy of the universe is multiplying in the hypothetical inflation is logically inconsistent, because we have no experimental evidence that energy can get multiplied. The inflation is against the first law of thermodynamics.

In the past century, gravity was understood as the force produced directly by the matter, the idea was that universe must be finite. We can read in the article of Sir James Jeans in Nature back in 1943: "If, however, the distribution is uniform throughout the whole of space, then space must be finite; otherwise it would contain an infinite amount of matter, and the gravitational force from this would be infinite, which is contrary to the fact" [14].

NASA has measured that the universe has Euclidean shape and is infinite [9]. The idea of SQS being infinite does not mean that gravity should be infinite, as suggested by Sir James Jeans. Considering universal space is infinite there is no gravity force between the stellar objects that are on the infinite distance.

The energy of the infinite universe in the form of matter E_m and in the form of space energy E_S is infinite:

$$E_m + E_S = \infty \quad (3).$$

The human mind can only imagine a finite amount of matter and a finite amount of energy and finite space which is not the case with the universe. The universe is infinite by means of matter, energy, and volume. That's why is opportune we study the universe that is at a finite distance and we predict that the rest of the unobservable universe on the infinite distance is behaving in the same way as our observable universe.

Mass of every physical object in the universe diminishes the energy density of space, the variable energy density of space is carrying gravity that is the fundamental force of the universal dynamics. Defining gravitational energy negative, as done by Hawking and Guth, are questionable; energy is not positive, it is not negative, energy simply is, it cannot be created and it cannot be destroyed, it transforms continuously.

Einstein has proposed on his steady—state theory of the universe that matter is continuously created out of the universal space: "In the final part of the manuscript, Einstein proposes a physical mechanism to allow the density of matter remain constant in a universe of expanding radius - namely, the continuous formation of matter from empty space: "If one considers a physically bounded volume, particles of matter will be continually leaving it. For the density to remain constant, new particles of matter must be continually formed within that volume from space" [15]. How the matter is formed out of space Einstein did not explain. Both, Hawking's and Einstein's solution for how matter appears in the universe are pure theoretical speculations. In our model appearance of matter in the universe is not questionable. In AGNs' matter is constantly disintegrating in elementary particles that are fresh energy for matter formation.

5. Universe is eternal and non-created

In the cosmology model proposed in this article the energy density of SQS in interstellar space has a value of Planck energy density $\rho_{EP} = 4.64 \cdot 10^{113} Jm^{-3}$. Every stellar object is diminishing energy density of SQS in its centre exactly for the amount of its mass m and energy E accordingly to the Eq. (1). Let's see the values of SQS energy density in the centre of some stellar objects on the table below:

Table 1. Comparation values of the energy density of space with respect to the centre of indicated objects.

Centre of objects	
	$\rho_{EP} = 4.64 \cdot 10^{113} Jm^{-3}$
Black hole with mass of the Sun	$\rho_{EP} - 1.58 \cdot 10^{36} Jm^{-3}$
Earth	$\rho_{EP} - 4.94 \cdot 10^{20} Jm^{-3}$
Moon	$\rho_{EP} - 3.00 \cdot 10^{20} Jm^{-3}$
Sun	$\rho_{EP} - 1.26 \cdot 10^{20} Jm^{-3}$

In the centre of a black hole with the mass of the Sun and corresponded Schwarzschild radius $r_{Sch} = 3 \cdot 10^3 m$, the minimal energy density of SQS is for the order of 10^{16} lower than in the centre of the Sun. Because of this special physical circumstance atoms become unstable. In the huge black holes in the centre of AGNs matter is falling apart into elementary particles that form jets. Black holes in the centre of galaxies are throwing these jets into intergalactic space. These jets are fresh energy for new stars formation; black holes are rejuvenating systems of the universe [16]. AGNs in the centres of galaxies are keeping entropy of the universe constant: "old" matter is transformed into "fresh" energy in the form of elementary particles (Figure 2).

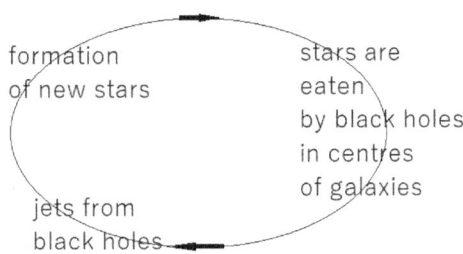

Figure 3: Energy circulation in the universe is permanent.

This process did not start and will never end, it is in permanent transformation. There was no creation of the energy of the universe and there will be no destruction of the energy. An increase of matter entropy in the universe is only a partial process that does not influence the total entropy of the universe that is constant. In AGN-s the universe is rejuvenating itself.

There is a strong astronomical evidence that the star HD 140283 has an age of 14.27 billion years [17] that is a new difficulty for Big Bang model according to which the age of the universe is calculated by about 13.7 billion years. This astronomical observation is another puzzle Big Bang cosmology should solve.

The main misunderstanding of Big Bang cosmology is that the universe has started in some remote physical past. We prove in this article there is no physical past. Universal changes are irreversible and run into time-invariant space. Universe has no physical past. Humans are thinking about the universal past as something real albeit it is non-existent. The only existent universe is the one we can observe. Cosmology's and astronomy's models about the physical past of the universe are not well understood. For example, we have a detailed model of how the Solar system was developed. But it was no developed in some physical past. It was developed in time-invariant space. Time-invariance of space we experience as what Einstein called NOW. Albert Einstein said: "...there is something essential about the NOW which is just outside the realm of science. People like us, who believe in physics, know that the distinction between the past, present, and future is only a stubbornly persistent illusion. Time has no independent existence apart from the order of the events by which we measure it" [18]. Considering NOW as the only existent physical reality means that the cosmological principle is time-invariant too. The universe as we observe it today is its real face. The initial explosion, inflation, and recombination are non-observed and remain theoretical speculations. We develop a cosmology model without the beginning of the universe, the problem of creation is solved. Penrose and Gurzadyan's "Conformal cyclic cosmology" (CCC) model also suggest that the universe is non-created, eternal, and in the permanent cyclic transformation [19]. CCC cosmology is accepting the inflation period that the "CPT – Symmetric universe" model [20] is denying.

CPT model predicts that before the big explosion there was an anti-universe in some negative time [20]. We categorically exclude that universe could exist in some negative time or could exist in some positive time. CCC cosmology model and CPT – Symmetric universe model weak points are that both models predict some events in the past that were never observed directly, their existence is questionable. The time-invariant eternal universe (TIEU) model is advanced in the sense it is based only on astronomical observations; it has no theoretical speculations about some past events in some remote physical past that does not exist. TIEU is based on the astronomical observations of the existing observable universe.

6. Conclusions

Universal space is time-invariant. This means that the cosmological principle too is-time-invariant. The universe as we see it today has the same face ever and forever. Humans have a huge power of imagination. In cosmology, this imagination went over the rational limits. Our research shows that the discrepancies between the Big Bang model and the obtained astronomical data are so huge that the Big Bang cosmology can be regarded as outdated. Seeing the universe as a system that exists in some physical time seems not right. The universe runs in the time-invariant superfluid quantum space. AGNs are rejuvenating systems of the eternal universe.

References:

1. Fiscaletti, D., & Šorli, A.S., Quantum Relativity: Variable Energy Density of Quantum Vacuum as the Origin of Mass, Gravity and the Quantum Behaviour, *Ukrainian Journal of*

Physics, 63(7), 623 (2018).
https://doi.org/10.15407/ujpe63.7.623.

2. Sbitnev, V.I. Hydrodynamics of superfluid quantum space: de Broglie interpretation of the quantum mechanics. *Quantum Stud.: Math. Found.* **5,** 257–271 (2018). **https://doi.org/10.1007/s40509-017-0116-z.**

3. Fiscaletti, D., Sorli, A.S. Perspectives of the Numerical Order of Material Changes in Timeless Approaches in Physics. *Found Phys* **45**, 105-133 (2015). **https://doi.org/10.1007/s10701-014-9840-y.**

4. Šorli, A., & Čelan, Štefan. (2020). The End of Space-time: Physics-Mathematics. *International Journal of Fundamental Physical Science*, *10*(4), 31-34. **https://doi.org/10.14331/ijfps.2020.330139.**

5. Šorli, A.S. Mass–Energy Equivalence Extension onto a Superfluid Quantum Vacuum. *Sci Rep* **9,** 11737 (2019). **https://doi.org/10.1038/s41598-019-48018-2.**

6. **Sorli, A.S. & Čelan Š. Superfluid quantum space as the unified field theory https://www.preprints.org/manuscript/202011.0460/v1** (sent to World Scientific)

7. Hirshfeld, Alan (2001). *Parallax: The Race to Measure the Cosmos*. New York: Henry Holt. **ISBN 978-0-8050-7133-7**.

8. Hartle, J.B., and Hawking, S.W. Wave function of the Universe. *Phys. Rev. D*, **28**, 2960 (1983). **https://doi.org/10.1103/PhysRevD.28.2960.**

9. NASA. **https://wmap.gsfc.nasa.gov/universe/uni_shape.html** (2014).

10. Zwicky, F. **On the Redshift of Spectral Lines Through Interstellar Space**. *Proceedings of the National Academy of Sciences*, **15**(10), 773-779 (1929). **doi:10.1073/pnas.15.10.773**. **PMC 522555**.

11. Jones, A.W., Lasenby, A.N. The Cosmic Microwave Background. *Living Rev. Relativ.*, **1**, 11 (1998). **https://doi.org/10.12942/lrr-1998-11.**

12. Fiscaletti, D., & Sorli, A.S. A Three-Dimensional Non-Local Quantum Vacuum as the Origin of Photons. *Ukrainian Journal of Physics*, **65**(2), 106 (2020). **https://doi.org/10.15407/ujpe65.2.106.**

13. **Guth, A., The Inflatory Universe.** *The Beamline*, **27, 14 (1997).**
    https://ned.ipac.caltech.edu/level5/Guth/Guth3.html.

14. **Sir James Jeans, The Structure of the Universe,** *NATURE*, **151, 190-192 (1943).**
    https://www.nature.com/articles/151490a0.pdf.

15. O'Raifeartaigh, C., McCann, B., Nahm, W. *et al.* Einstein's steady-state theory: an abandoned model of the cosmos. *EPJ H* **39,** 353–367 (2014). **https://doi.org/10.1140/epjh/e2014-50011-x.**

16. Sorli, A.S. Čelan, Š., Quantum mechanism of AGN's jets **https://www.preprints.org/manuscript/202011.0581/v1** sent to HICARI

17. **VandenBerg**, **D.A.**, **Bond**, **H.E.**, **Nelan**, **E.P.**, **Nissen**, **P.E.**, **Schaefer**, **G.H.**, **Harmer**, D. Three Ancient Halo Subgiants: Precise Parallaxes, Compositions, Ages, and Implications for Globular Clusters (2014). **https://arxiv.org/abs/1407.7591.**

18. Fiscaletti, D. Sorli, A.S. The Infinite History of Now: A Timeless Background for Contemporary Physics, ISBN-10: 1631172832, Nova publishing (2014).

19. Gurzadyan, V.G., Penrose, R. On CCC-predicted concentric low-variance circles in the CMB sky. *Eur. Phys. J. Plus* **128,** 22 (2013). **https://doi.org/10.1140/epjp/i2013-13022-4.**

20. Latham Boyle, Kieran Finn, and Neil Turok, CPT-Symmetric Universe, Phys. Rev. Lett. 121, 251301 – Published 20 December 2018
https://doi.org/10.1103/PhysRevLett.121.251301.

Cosmo-biology
Multiverse, Life and Consciousness

Abstract

The evolution of life on the planet Earth is happening primarily in the universe and secondary on the Earth. We will examine in this article evolution of life as the cosmic phenomena. In our model multidimensional time-invariant superfluid quantum space that is the fundamental arena of the universe and represents about 95% of the energy in the universe has stable entropy. The increase of entropy happens only by about 5% of the energy in the universe that is in the form of matter. The evolution of life in our model is a process of matter organization into living systems that tends to develop towards the constant entropy of the time-invariant multidimensional quantum space. This process runs in the entire universe. The development of life into intelligent organisms is the universal process running throughout the entire universe.

Keywords: cosmology, life, superfluid quantum space, consciousness.

1. Introduction

The result of several pieces of research is that the superfluid quantum vacuum also named superfluid quantum space (SQS) is the physical origin of the universal space, the fundamental arena of the universe [1,2,3,4,5]. Superfluid quantum space (SQS) has a general n-dimensional complex structure $\mathbb{C}^n$; every point of it has complex coordinates:

$$z_i = x_i + i y_i \quad (1).$$

(x_i, y_i) $(i = 1, \ldots, n)$ is an ordered n-uple of real numbers $((x_i, y_i) \in \mathbb{R}^n)$; for the purpose of this paper, we consider its subset $\mathbb{C}^4$ where all elementary particles are different structures of $\mathbb{C}^4$SQS and have four complex dimensions z_i "[4].

Elementary particles proton, electron, and photon are 4-dimensional structures of the $\mathbb{C}^4$SQS and have according to the existing quantum theory almost infinite lifetime. Sbitnev proposal is that elementary particles are different vortex structures of superfluid quantum space [6]. As $\mathbb{C}^4$SQS has stable entropy, proton, electron and photon have stable entropy. 5% of the energy in the universe is in form of matter composed out of atoms that are 3-dimensional and 95% is in the form of the $\mathbb{C}^4$SQS that is 4-dimensional and has stable entropy.

Multiverse is in dynamic equilibrium. In AGNs matter is falling apart into elementary particles that form huge jets. These jets are the "raw material" for the formation of the new stars. This process of matter transformation in elementary particles is continued. The universe is a non-created system in a permanent dynamic equilibrium [4,5]. The evolution of life in the universe is an intrinsic tendency of 3D matter to develop into systems (living organisms) that tend to develop towards the constant entropy of $\mathbb{C}^n$SQS.

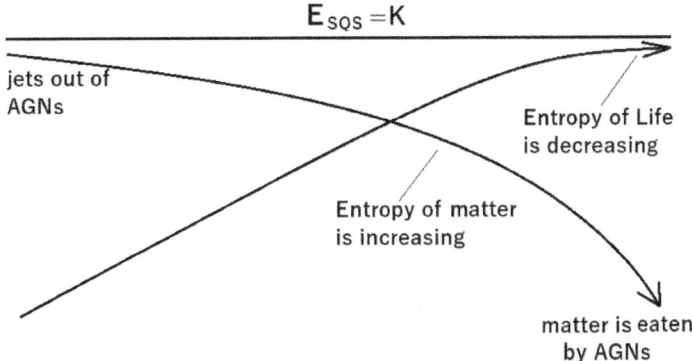

Figure 1: Life is developing towards the constant entropy of $\mathbb{C}^n SQS$

The entropy of matter in the universe is continuously increasing. The entropy of life is continuously decreasing. The entropy of $S_{\mathbb{C}^n SQS}$ is constant:

$$S_{\mathbb{C}^n SQS} = K \quad (2).$$

2. Quantum mechanics of life negentropy

Fritz Popp's and Cohen's research has shown that a living organism has a coherent electromagnetic field that plays an essential role in the organism's function [7]. Electromagnetic fields are carried by the complex four-dimensional superfluid quantum space $\mathbb{C}^4 SQS$. A living organism has an atomic and molecular layer entropic layer and electromagnetic layer that is negentropic. The negentropic layer is the software and the entropic layer is the hardware. Our proposal in this article is that life is an "orchestra" of the higher dimensional layers of SQS. Consciousness in our model is the energy E_c of the photon of the n-dimensional layer of SQS; its frequency tends to the infinite:

$$E_c = v_{\to \infty} h \quad (3),$$

where v is photon frequency and h is a Planck constant. Consciousness is governing life via lover dimensional SQS by the pilot photons. Biophotons are studied in detail by Popp and Cohen [7]. Photons have a spin. Left spin, we can take like 1, and right spin we can take like 0. When a photon is passing the microtubule, it passes the information via its spin. $\mathbb{C}^4$SQS photons have 4 bites of the information. They are getting information from higher dimensional SQS photons and are passing it to the microtubule [8]. The equation for the information increases in higher dimensional layers of SQS is following:

$$C_k(n) = \frac{n!}{(r!(n-r)!)} \quad (4)$$

where n is the number of SQS dimensionality, and $r = 3$ because microtubules are 3-dimensional. A four-dimensional photon can carry 4 bit of information: $[X_1, X_2, X_3]$, $[X_2, X_3, X_4]$, $[X_1, X_2, X_4]$, $[X_1, X_3, X_4]$ and transfers it to the microtubules. Microtubules are 3D and can receive only information in the 3D form.

Table 1. Information density in higher dimensions of $\mathbb{C}^n$SQS.

$\mathbb{C}^4$SQS	4 bit
$\mathbb{C}^5$SQS	10 bit
$\mathbb{C}^6$SQS	20 bit
$\mathbb{C}^7$SQS	35 bit
$\mathbb{C}^8$SQS	56 bit
$\mathbb{C}^9$SQS	84 bit
$\mathbb{C}^{10}$SQS	120 bit
$\mathbb{C}^{100}$SQS	161700 bit
$\mathbb{C}^n$SQS	∞ bit

In n-dimensional SQS the amount of information is infinite. Seems, life and the entire universe are functioning via binary logic and binary transfer of information. That's why we managed the immense development of computers; we discovered the mechanisms of information storage and transfer that are universal. The numbers sequence 4,10,20,35,56,84,120.......is a tetrahedral sequence of numbers, also called triangular pyramidal numbers. It is interesting that several molecules have a tetrahedral structure [9,10]. Tegmark is proposing that the entire universe is a mathematical structure [11]. Comparing Tegmark's proposal our model is moderate and proposes that the entire three-dimensional universe is built accordingly to the mathematical structures that have their information basis in the higher-dimensional layers of SQS.

3. Materials and methods

Research done between 1987-90 has confirmed that the presence of the higher dimensional layers of SQS in a living organism minimally increases its weight. The weight of one gram living worms is about one-million-part bigger of the weight of same dead worms:

$$Fg = Fg_{matter} + Fg_{life} \qquad (5).$$

in units: $1g = 0{,}999999\ g + 0{,}000001g$

The mass m of living worms and the mass m of dead worms are the same because in both masses we have the same atoms. Only their molecular composition after poisoning with formaldehyde is different. A living organism has more energy than the same dead organism. Its energy is following:

$$E_{life} = mc^2 + E_{\mathbb{C}^n SQS} \qquad (7),$$

where m is the mass of the living organism and $E_{\mathbb{C}^n SQS}$ are higher-dimensional energies of $\mathbb{C}^n SQS$ that are present in the living organism. The presence of higher-dimensional energies of $\mathbb{C}^n SQS$ in living organism minimally increases its weight [12].

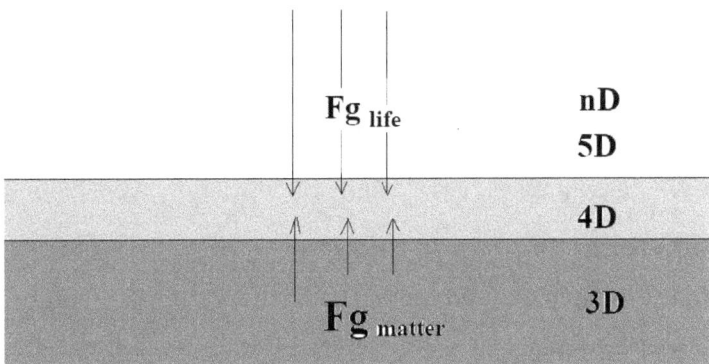

Figure 2: Gravity of matter (Fg_{matter}) and gravity of life (Fg_{life})

Gravity force is the result of the diminished energy density of $\mathbb{C}^4$SQS because of the presence of the physical object [4]. The presence of higher dimensional layers of SQS also diminishes the energy density of $\mathbb{C}^4$SQS and so gravity force is minimally increased. This is so called "gravity force of life" (Fg_{life}), see figure above.

We could say that so-called "subtle energies" as "Prana" or ""QI" energy and consciousness have some minimal weight [12]. So called "orthodox science" is strictly denying existence of reality that reaches beyond the electromagnetism. We think this approach will not give us any progress. As Nicola Tesla said: "The day science begins to study non-physical phenomena, it will make more progress in one decade than in all the previous centuries of its existence".

Slawinski has measured bio-photon radiation at the time of death of the organism and increases from 10 to 100 times [13]. This confirms that at the time of death the four-dimensional layer of SQS that represents the coherent electromagnetic field discover by Popp and Cohen is falling apart and this causes increased bio-photons radiation.

4. Evolution of life, order, disorder and randomness

We take a "fair coin" and we throw it. We have a 50% possibility to get the "upper side" and a 50% possibility to get a "downside" of the coin. We take a "fair dice" with six numbers. We throw it and we have a 16.66 % possibility to get number six.

We take two fair dices, we place them on the plate so that they both have number six on the upper surface and we throw them. We use the equation (4) to get the number of possibilities. Number n is 12 because we have 6+6 surfaces, and number k is 2. Throwing two dices we can get 66 different combinations. This means that the possibility that both dices will have at next throw number six on the upper surface is 1.56%. Now we take 10 dices and we place them so that all have the number six on the upper surface. Number n is 60, and number k is 10. At the next throw, we have 75394027566 different possibilities. Possibility that all dices will have at next throw number six on upper surface is $1.326 \cdot 10^{-9}$ %. At 100 dices number n is 600, and number k is 100. At the next throw, we have $1.111 \cdot 10^{116}$ different possibilities. Possibility that all dices will have at next throw number six on upper surface is $9 \cdot 10^{-115}$ %.

Random hitting of dices increases the disorder of the system. A living organism's order is extremely bigger than the order of the system of 100 dices. Life is regarding the geological environment extremely high organized system. Longo and Montévil have proposed that randomness increases order in biological evolution

[14]. The calculations above confirm that the idea that randomness is the cause of biological evolution seems unacceptable.

Penrose and Hameroff have proposed consciousness as the core of life evolution. They have created orchestrated objective reduction theory (Orch OR), which sees life and consciousness as phenomena that are deeply related to the structures of the universe: "The DP (Diósi–Penrose) form of OR is related to the fundamentals of quantum mechanics and space-time geometry, so Orch OR suggests that there is a connection between the brain's biomolecular processes and the basic structure of the universe" [15]. We have replaced space-time model with the time-invariant model. Seeing consciousness as something that appears in time is outdated. Linear phycological time "past-present-future" exists only in the human brain and consciousness is far beyond the brain and phycological time [16]. The universe is time-invariant, time as duration enters existence when measured by the observer [5]. In our model evolution of life has its information basis in the higher dimensional layers of SQS. Entire universe is existing in a time-invariant SQS, everything in the universe is entangled via time-invariant SQS [4]. In $\mathbb{C}^4$SQS information transfer is of the light speed. In $\mathbb{C}^n$SQS information transfer is immediate. $\mathbb{C}^n$SQS is the medium of EPR-type entanglement. This is because the frequency of the photon in $\mathbb{C}^n$SQS that represent consciousness tends to infinite value and its wavelength tends to zero. Its velocity tends to zero ($v_{\to \infty}, \lambda_{\to 0}, v = v$ $\lambda = \to 0$). Consciousness is the carrier of the immediate information transfer. Back in 2014, Max Tegmark published an article where he discussed that consciousness could be understood as a state of mater

[17]. In the $\mathbb{C}^n$SQS model, all that exists in the universe is energy. Matter, electromagnetic energy, and consciousness are different aspects of the same energy. There is no need to think that matter is primary and consciousness is a state of matter or that consciousness is primary and the matter is its manifestation. They are both coexistent forms of the same energy. In $\mathbb{C}^n$SQS model dichotomy matter/consciousness is solved. Energies of all layers of $\mathbb{C}^n$SQS are interwoven. They are one organism and seeing them separate seems wrong.

Einstein and Bohm were not in the favour of the idea that the universe is a random phenomenon with no order. Einstein has proposed "hidden variables" to explain the EPR-type experiments, Bohm has proposed "implicate order of the universe", a model that proposes the universe is an intelligent system [18]. There is a deep ontological similarity between Einstein's, Bohm's, and our view. In our model universe is governed by consciousness and we humans have to search consciousness experientially in order to be able to follow cosmic laws and build human society accordingly [19].

Self-organization is today recognized as a valid principle in developmental biology [20,21]. It is well recognized that life is organizing itself. The mistake is to believe that this principle is ruling the development of life. No principle can rule a given process. A given principle in order to be real needs discovery of its physical origin. The principle of self-organization needs experimental verification. Our experiment "life-dead weight difference" proves that some higher dimensional type of $\mathbb{C}^n$SQS energies is present in the living organism. These higher-dimensional energies of $\mathbb{C}^n$SQS are the physical origin of self-organization. It makes no sense to see living organisms as an isolated system. Life is deeply related to the $\mathbb{C}^n$SQS.

Organic molecules have been found in the interstellar medium [22]. In our model interstellar medium is the $\mathbb{C}^n$SQS. Molecules in interstellar areas have a tendency of self-organization because information of life is encoded in higher-dimensional layers of SQS. On the planets that are similar to the planet Earth, life has developed in intelligent beings. In our universe, there are many planes similar to our planet Earth where life could develop [23,24]. Cosmo-biology is reaching beyond anthropocentrism and geo-centrism. We are not the centre of the universe. The evolution of life on Earth is the consistent part of a universal process that runs throughout the entire universe. Considerable progress in this direction is done by Meyer, Jerman, Melkikh, and Sbitnev [25]. Their model has several parallels with our model; the main difference is that they see the evolution of life in some physical time and we see time "past-present-future" only as a psychological reality. In our model universe and life are developing in the time-invariant $\mathbb{C}^n$SQS.

5. Conclusions

The universe is the main system in which all other systems exist. It is opportune to approach life as a universal process. Astrobiology is searching for extra-terrestrial life. Cosmo-biology is building a model where the development of life is an integral part of the universal dynamics. Higher-dimensional layers of SQS are the cosmic reservoir of information for the development of life.

References:

1. Sbitnev, V.I. Hydrodynamics of superfluid quantum space: de Broglie interpretation of the quantum mechanics. *Quantum Stud.: Math. Found.* **5,** 257–271 (2018). **https://doi.org/10.1007/s40509-017-0116-z**.

2. A. Šorli, Š. Čelan, Advances of Relativity Theory, Physics Essays (accepted in publication 15.3.2021, online in June 2021)

3. Amrit S. Sorli, Stefan Čelan, **Schwarzschild energy density of superfluid quantum space and mechanism of AGNs' jets**, Advanced Studies in Theoretical Physics, Vol. 15, no. 1, 9-17. (2021) **https://doi.org/10.12988/astp.2021.91506**.

4. **Šorli, A.S. & Čelan Š. Time-invariant superfluid quantum space as the unified field theory, Reports in Advances of Physical Sciences**, World Scientific (**accepted for publication 31.2.2021, in press**),

5. A. Šorli, Š. Čelan, A.Brzo, Multiverse in Dynamic Equilibrium (will be published in June)

6. Sbitnev, V.I. Hydrodynamics of the Physical Vacuum: II. Vorticity Dynamics. *Found Phys* **46,** 1238–1252 (2016). **https://doi.org/10.1007/s10701-015-9985-3**.

7. Cohen S and Popp FA. Biophoton emission of the human body. Journal of Photochemistry and Photobiology B: Biology 1997; 40(2): 187-189. **https://doi.org/10.1016/S1011-1344(97)00050-X**.

8. Amrit Sorli, Uros Dobnikar, Davide Fiscaletti, Vlad Koroli, Advanced Relativity: Multidimensionality of Consciousness and Mind, Origin of Life, PSI Phenomena, *Volume 15, No 2 (2017)* DOI: **10.14704/nq.2017.15.2.1020**.

9. Mason, P. E.; Brady, J. W. (2007). ""Tetrahedrality" and the Relationship between Collective Structure and Radial Distribution Functions in Liquid Water". *J. Phys. Chem. B*. **111** (20): 5669–5679. **doi:10.1021/jp068581n**.

10. **Wiberg, Kenneth B.** (1984). "Inverted geometries at carbon". *Acc. Chem. Res.* **17** (11): 379–386. **doi:10.1021/ar00107a001**.

11. Tegmark, Max (2008). "The Mathematical Universe". *Foundations of Physics*. **38** (2): 101–150. **doi:10.1007/s10701-007-9186-9**.

12. Amrit Sorli, Santanu Kumar Patro, Davide Fiscaletti, Unified Field Theory Based on Bijective Methodology, NeuroQuantology | December 2017 | Volume 15 | Issue 4 | Page 37-44 | doi:10.14704/nq.2017.15.4.1106.

13. Slawinski J. Photon emission from perturbed and dying organisms: biomedical perspectives. Complementary Medicine Research 2005;12(2):90- 95. **https://doi.org/10.1159/000083971**.

14. Longo G., Montévil M. (2012) Randomness Increases Order in Biological Evolution. In: Dinneen M.J., Khoussainov B., Nies A. (eds) Computation, Physics and Beyond. WTCS 2012. Lecture Notes in Computer Science, vol 7160. Springer, Berlin, Heidelberg. **https://doi.org/10.1007/978-3-642-27654-5_22**.

15. Stuart Hameroff, Roger Penrose, Consciousness in the universe A review of the 'Orch OR' theory, Physics of Life Reviews 11 (2014) 39–78 **https://doi.org/10.1016/j.plrev.2013.08.002**.

16. Šorli, A., & Čelan, Štefan. (2020). Einstein's misunderstanding of time in time-invariant universe: Physics-Mathematics. *International Journal of Fundamental Physical Science* (accepted for publication 3.2.2021, in press, **expected publication March 2021**)

17. Max Tegmark, Consciousness as a state of matter, Chaos, Solitons & Fractals, Volume 76 (2015) Pages 238-270, **https://doi.org/10.1016/j.chaos.2015.03.014**.

18. David Bohm: *Wholeness and the Implicate Order*, Routledge, 1980 (**ISBN 0-203-99515-5**).

19. Sorli A., Kaufman S., Experiential Consciousness Research, NeuroQuantology 2018; 16(3) 10.14704/nq.2018.16.3.1132.

20. Karsenti, E. Self-organization in cell biology: a brief history. *Nat Rev Mol Cell Biol* **9,** 255–262 (2008). **https://doi.org/10.1038/nrm2357**.

21. **Wedlich-Söldner Roland** and **Betz Timo** (2018) Self-organization: the fundament of cell biologyPhil. Trans. R. Soc. B37320170103 **http://doi.org/10.1098/rstb.2017.0103**.

22. Pascale Ehrenfreund and Steven B. Charnley, **Organic Molecules in the Interstellar Medium, Comets, and Meteorites: A Voyage from Dark Clouds to the Early Earth**, Annual Review of Astronomy and Astrophysics 2000 38:1, 427-483 **https://www.annualreviews.org/doi/10.1146/annurev.astro.38.1.427**.

23. **Dirk Schulze-Makuch, René Heller, Edward Guinan**, In Search for a Planet Better than Earth: Top Contenders for a Superhabitable World, **AstrobiologyVol. 20, No. 12** (2020) **https://doi.org/10.1089/ast.2019.2161**.

24. The Search for Life, NASA (2020) **https://exoplanets.nasa.gov/search-for-life/habitable-zone/**.

25. Meijer, Dirk & Jerman, Igor & Melkikh, Alexey & Sbitnev, Valeriy. (2019). Consciousness in the Universe is Tuned by a Musical Master Code: A Hydrodynamic Superfluid Quantum Space Guides a Conformal Mental Attribute of Reality. The Hard Problem in Consciousness Studies Revisited.

Mass Gap Problem and Planck constant

Amrit S. Šorli
Bijective Physics Institute, Slovenija
sorli.bijective.physics@gmail.com
https://orcid.org/0000-0001-6711-4844

Štefan Čelan
Scientific research centre Bistra, Slovenija
stefan.celan@bistra.si
https://orcid.org/0000-0003-3646-1469

Abstract

The mass gap problem is about defining the constant that defines the minimal excitation of the vacuum. From the point of quantum mechanics, Planck's constant is defining the minimal possible excitation of the vacuum. The mass gap problem can be solved in the frame of quantum mechanics by the formulation of the photon's mass accordingly to the Planck-Einstein relation.

Keywords: Mass Gap Problem, Planck-Einstein relation, superfluid quantum space, bijectivity, falsifiability.

1. Introduction

To solve the Yang-Mills Mass Gap Problem [1] we have to understand well what mass is. We propose a bijective solution for the mass-gap problem where every element in the model of physical reality has exactly one correspondent model in physical reality. The model of physical reality is set Y and physical reality is set X. Every element in set Y has exactly one element in the set X. For example, element mass m_x in the physical universe has correspondent element mass m_y in the model of the universe. They are related by the bijective function of set theory:

$$f: m_x \to m_y \quad (1).$$

Einstein equation $E = mc^2$ has bijective correspondence with the physical universe:

$$E_x = m_x c_x^2$$
$$E_Y = m_y c_y^2 \quad (2).$$

Every object with mass m is existing in space we call today a superfluid quantum space (SQS). SQS is not "empty", it is the fundamental energy of the universe. Elementary particles are different structures of SQS [2]. In the space-time model of Special Relativity, the fourth coordinate X_4 is imaginary, X_1, X_2, X_3 are real coordinates. In the time-invariant space model, the 4th coordinate Z_4 is a complex coordinate as the other three coordinates Z_1, Z_2, Z_3: "Time-invariant superfluid quantum space (SQS) has a general n-dimensional complex structure $\mathbb{C}^n$; every point of it has complex coordinates:

$$z_i = x_i + i y_i \quad (3).$$

(x_i, y_i) $(i = 1, ..., n)$ is an ordered n-uple of real numbers $((x_i, y_i) \in \mathbb{R}^n)$; for the purpose of this paper, we consider its subset $\mathbb{C}^4$ where all elementary particles are different structures of $\mathbb{C}^4$-SQS and have four complex dimensions z_i "[2]. The model of the superfluid quantum space is close to the model of 4D superfluid quantum space-time, also named "superfluid ether" which uses quaternions as its mathematical background [3].

The idea of "empty" space as something real is harming physics for more than 100 years. It is time we demolish the idea of empty space and introduce ether back into physics. Michelson-Morley's experiment has given null results because it was carried out by the proposition that Earth is moving through a stationary ether. Ether around the physical object is moving and rotating with the physical object. Masanory research suggested that on the distance of 20000 km above the Earth ether drift could be measured: "The satellites in the higher orbit (in the yellow region) have a possibility to detect the ether-drift. The evidence of the ether-drift can be proven by the fact that the ECI coordinate system does not work well. Of course, these experiments have not been carried out yet. The discussions of the ether-drift and frame-dragging were carried out more than 100 years ago. I have not carried out any calculation of the height of the frame-dragging using the theory of general relativity. At this stage, I consider that the height of the ether-drift detected is more than 20,000 km from the ground level" [4].

Mass m is in physics an element with the attribute of energy E. It is false to think that this element mass can exist in space that has no attribute of energy E. The equation (4) below is false:

$$f: empti\ space_x \rightarrow empty\ space_y \quad (4).$$

2. Mass Gap Problem from the perspective of mass-energy principle extension on SQS and Planck-Einstein relation

Mass m of a given physical object is related to the energy of SQS accordingly to the well-known physical law of homogenous distribution of energy. Every physical system tends that the energy of the system is distributed in a homogeneous way; because of this every physical object with mass m is diminishing the energy density of SQS exactly for the amount of its energy E:

$$\frac{E}{c^2} = m = \frac{(\rho_{Emax} - \rho_{Emin})V}{c^2} \quad (5)\ [2],$$

where ρ_{Emax} is the energy density of the SQS infinitely far away from the physical object surface, ρ_{Emin} is the energy density of the SQS in the centre of a given physical object and V is the volume of the object. In this way energy density of SQS remains uniform. In the Newtonian perspective, the area of space with a higher density is asserting a given pressure towards the area of space with a lower density. From a quantum physics perspective, universal space has the vector orientation towards lower energy density, or in space are quantum fluctuations towards lover energy density of space.

Equation (5) is valid from the scale of the proton to the scale of AGN. It can describe the process of the formation of the jets in the AGNs. The energy density of SQS there is so low that atoms become unstable and form jets that are spreading into intergalactic space [5]. Equation (5) is the extension of the mass-energy equivalence principle on the universal space that has its origin in time-invariant superfluid quantum space SQS and can solve the mass gap problem. The variable energy density of SQS is giving the origin to the inertial mass m_i and gravitational mass m_g as follows in Eq. (6) below:

$$m_i = m_g = \frac{(\rho_{Emax} - \rho_{Emin})V}{c^2} \qquad (6),$$

where ρ_{Emax} is the energy density of the space in interstellar space; ρ_{Emin} is the energy density of the space in the centre of the proton, and V is the volume of the proton.

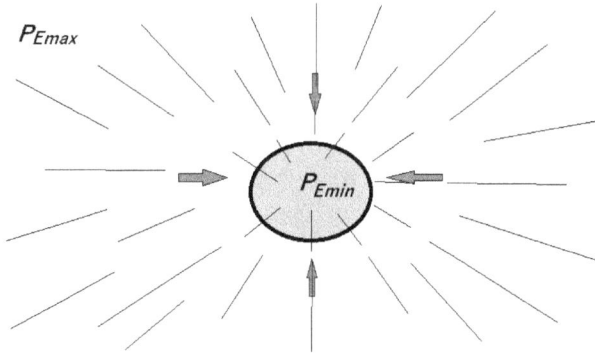

Figure 1: Proton inertial mass and proton gravitational mass have the same origin in variable energy density of space

Equation (6) shows that inertia and gravity are the results of the dynamics between given mass m and a variable energy density of SQS. It is false to think that inertial mass m_i is the same phenomenon as the rest mass m_0 as the amount of energy E Equation (7) below is false:

$$\frac{E}{c^2} = m_0 = m_i \qquad (7).$$

We can combine two fundamental equations $E = mc^2$ and equation $E = hv$ and we get:

$$E = mc^2 = hv \qquad (8)$$

$$m = \frac{hv}{c^2} \qquad (9).$$

Photon is massless in the sense that it has no inertial mass m_i. But photon has energy E and so it has correspondent mass m. Equation (9) is showing the mass of the photon related to its frequency. Combining (5) and (9) we get:

$$m = \frac{(\rho_{Emax} - \rho_{Emin})V}{c^2} = \frac{hv}{c^2} \qquad (10).$$

Out of (10) follows:

$$h = \frac{(\rho_{Emax} - \rho_{Emin})V}{v} \qquad (11).$$

The space energy density difference $\rho_{Emax} - \rho_{Emin}$ we can express as $\Delta_{\rho E} = \rho_{Emax} - \rho_{Emin}$ and we get:

$$h = \frac{\Delta_{\rho E} V}{v} \qquad (12).$$

Eq. (12) shows that the value $\frac{\Delta_{\rho E}}{v}$ is constant. When frequency v is increasing delta energy density $\Delta_{\rho E}$ is also increasing:

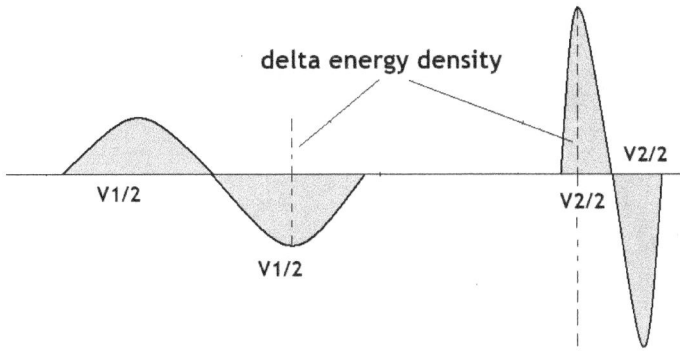

Figure 2. An increase in frequency increases delta energy density and keeps the same volume.

The volume of the photon is introduced as an approximation to describe how much of the volume of the SQS is in excitation when we have a single photon. Frequency and wavelength are related by the formula $v = \frac{c}{\lambda}$, where c is the light speed. When the frequency is growing, the wavelength is getting smaller. This ideal volume V of the photon is independent of the photon frequency and its wavelength; see figure 2. where volume V1 of the photon with the low frequency and the volume V2 of the photon with high frequency are equal: V1 = V2. Eq. (12) is bijective with the physical reality. Bijectivity assures falsifiability, Eq. (12) is falsifiable.

For Yang-Mills Mass Gap Solution is required: "It must have a "mass gap;" namely there must be some constant $\Delta > 0$ such that every excitation of the space has energy at least Δ " [1]. Eq. (12) shows this constant $\Delta > 0$ is Planck constant h. Namely, every particle can be seen as the excitation of the $\mathbb{C}^4$-SQS and is defined by the difference of vacuum pressure Δp_E, volume V and frequency v of a given particle. These three elements are defining the Planck constant h. A given excitation of $\mathbb{C}^4$-SQS is producing a given elementary particle with energy E. We know the origin of the Planck constant (Eq. 12) and we see that the minimal "mass gap" in the excitation of SQS is defined by the Planck constant h. In physical terms "mass gap" would be better named as "mass-energy gap", because Einstein told us already that mass and energy are made out of the same "stuff". What's this "stuff" is being clearly explained by Erving Schrodinger who also regarded space as the fundamental energy of the universe: "What we observe as material bodies and forces are nothing but shapes and variations in the structure of space". Reintroduction of the $\mathbb{C}^4$-SQS model and of the mass-energy equivalence extension on the $\mathbb{C}^4$-SQS are helpful tools to develop physics where physical objects are the energy structures of the medium in which they exist. The electric field is the excitation of the 4th layer of $\mathbb{C}^4$-SQS along the dimensions z_1, z_2, z_3. The magnetic field is the excitation of the 4th layer of $\mathbb{C}^4$-SQS along the dimensions z_2, z_3, z_4. Magnetic and electric fields have in common dimensions z_2, z_3. Photon is the four-dimensional wave of the 4th layer of $\mathbb{C}^4$-SQS. Time is the duration of photon motion in $\mathbb{C}^4$-SQS [2].

3. Conclusions

We proposed in this article the solution of the "mass gap" problem based on the Planck-Einstein relation that occurs in the superfluid quantum space $\mathbb{C}^4$-SQS. The result is that in the view of the mass-energy equivalence principle, the minimal change of "mass-energy" excitation and so the minimal mass gap in the superfluid quantum space $\mathbb{C}^4$-SQS and is defined by Planck constant h.

References:

1. Clay Mathematics Institute
   https://www.claymath.org/millennium-problems (2020)

2. **Šorli, A.S. & Čelan Š. Time-invariant Superfluid Quantum Space as the Unified Field Theory, Reports in Advances of Physical Sciences**, World Scientific, **accepted for publication 31.1.2021 (in press),**

3. Sbitnev, V. Quaternion Algebra on 4D Superfluid Quantum Space-Time: Can Dark Matter Be a Manifestation of the Superfluid Ether? UNIVERSE **2021**, 7, 32.
   https://doi.org/10.3390/universe7020032.

4. Masanori S., Gravitational effect on the refractive index: A hypothesis that the permittivity, ε0, and permeability, μ0 are dragged and modified by the gravity
   https://arxiv.org/vc/arxiv/papers/0704/0704.1942v3.pdf.

5. Amrit S. Sorli, Stefan Celan, **Schwarzschild energy density of superfluid quantum space and mechanism of AGNs' jets**, Advanced Studies in Theoretical Physics, Vol. 15, no. 1, 9-17. (2021) **https://doi.org/10.12988/astp.2021.91506**.

www.ingramcontent.com/pod-product-compliance
Lightning Source LLC
Chambersburg PA
CBHW050007230526
45465CB00003BB/1302